Thomas Nuttall, François André Michaux, John Jay Smith

The North American Sylva

Vol. 4

Thomas Nuttall, François André Michaux, John Jay Smith

The North American Sylva
Vol. 4

ISBN/EAN: 9783337306816

Printed in Europe, USA, Canada, Australia, Japan

Cover: Foto ©berggeist007 / pixelio.de

More available books at **www.hansebooks.com**

THE

North American Sylva;

OR, A DESCRIPTION OF THE

FOREST TREES

OF THE

UNITED STATES, CANADA, AND NOVA SCOTIA,

NOT DESCRIBED IN THE WORK OF

F. ANDREW MICHAUX,

AND CONTAINING ALL THE

FOREST TREES DISCOVERED IN THE ROCKY MOUNTAINS, THE TERRITORY OF OREGON, DOWN TO THE SHORES OF THE PACIFIC, AND INTO THE CONFINES OF CALIFORNIA, AS WELL AS IN VARIOUS PARTS OF THE UNITED STATES.

ILLUSTRATED BY 121 COLORED PLATES.

BY

THOMAS NUTTALL, F.L.S.

MEMBER OF THE AMERICAN PHILOSOPHICAL SOCIETY, AND OF THE ACADEMY OF NATURAL SCIENCES OF PHILADELPHIA, ETC. ETC. ETC.

THREE VOLUMES IN TWO.

VOL. I.

BEING THE FOURTH VOLUME OF MICHAUX AND NUTTALL'S NORTH AMERICAN SYLVA.

PHILADELPHIA:
PUBLISHED BY RICE, RUTTER & CO.,
No. 525 MINOR STREET.
1865.

Entered according to Act of Congress, in the year 1865, by

RICE, RUTTER & CO.,

In the Clerk's Office of the District Court of the United States for the Eastern District of Pennsylvania.

STEREOTYPED BY L. JOHNSON AND CO.
PHILADELPHIA.

COLLINS, PRINTER.

TO THE LATE

WILLIAM MACLURE, ESQ.

PRESIDENT OF THE ACADEMY OF NATURAL SCIENCES IN PHILADELPHIA, ETC. ETC.

AS A MEMENTO OF HIS ATTACHMENT TO, AND LIBERAL ENCOURAGEMENT OF, NATURAL SCIENCES IN NORTH AMERICA;

ALSO, TO

F. ANDREW MICHAUX,

MEMBER OF THE AMERICAN PHILOSOPHICAL SOCIETY, CORRESPONDENT OF THE INSTITUTE OF FRANCE, ETC. ETC.,

WHOSE NAME IS IDENTIFIED WITH THE HISTORY AND IMPORTANCE OF THE PRODUCTIONS OF THE NORTH AMERICAN FOREST,

This Work

IS MOST RESPECTFULLY DEDICATED BY

THE AUTHOR OF THIS SUPPLEMENT.

PREFACE.

THE FOREST TREES OF AMERICA being a subject of such great extent and importance, I felt, consequently, very diffident of undertaking their study, after what has been already done so well by my predecessor, M. MICHAUX. Yet, in offering a new edition of the AMERICAN SYLVA in English, it appeared requisite, in keeping pace with the progress of discovery, that all the forest trees of the extended dominion of the United States should, in some way or other, be included in the present publication; and, I confess, the magnitude of the task appeared, at first, sufficiently appalling, when we reflect on the vast territory now claimed by the United States. Beginning with the arctic limits of all arborescent vegetation, in the wilds of Canada, which we cannot with propriety exclude, forming as it does the boreal boundary of the North American forest, we then follow the extended shores of the Atlantic, until, toward the extremity of East Florida, and its keys or islands, we have attained the very confines of the tropical circle, and make a near approach to the island of Cuba and the Bahamas. Turning westward, we pass over the wide forests of the Mississippi, pursue the Western streams, through vast woodless plains, until we attain the long crests of the Rocky Mountains or Northern Andes. Here, in these alpine regions, we meet with a total change in the features of the forest: resiniferous evergreens, of the family of the Pines, now predominate, and attain the most gigantic dimensions. All the species (and they are numerous) have peculiar traits, and form so many curious and distinct species, of which little is yet known more than their botanical designation. Other remarkable forest trees, also imperfectly known, inhabit this great range of mountains, which continues uninterruptedly into the interior of Mexico in its southern course; while on the north, following the sources of the Missouri and the Oregon, and after thus dividing the waters which flow into the Atlantic and Pacific, it is at length merged in the "Shining Mountains," which send off their distant tributaries to the Arctic Ocean.

The plains of the Upper Platte, those of the Oregon and of Northern California, a region bereft of summer rains, forming extensive barren steppes, like those of Siberia, present no forests, scarcely an alluvial belt along the larger streams of sufficient magnitude to afford even fuel for the camp-fire of the wandering hunter or the erratic savage. The scanty driftwood borne down from the mountains, the low bitter bushes of the arid plain, even the dry ordure of the bison, is collected for fuel, and barely suffices to prepare a hasty meal for the passing traveller, who, urged by hunger and thirst, hurries over the desert, a region doomed to desolation, and, amid privations the most appalling, lives in the hope of again seeing forests and green fields in lieu of arid plains and bitter weeds, which tantalized our famished animals with the fallacious appearance of food, like the cast-away mariner raging with thirst, though surrounded with water as fatal to the longing appetite as poison.

Toward the shores of the Pacific, and on the banks of the Oregon, we again meet with the agreeable features of the forest:—

> "Majestic woods, of every vigorous green,
> Stage above stage, high waving o'er the hills,
> Or to the far horizon wide diffused,
> A boundless, deep immensity of shade."

Transported in idea to the border of the Hudson or the Delaware, we recline beneath the shade of venerable Oaks and spreading Maples; we see, as it were, fringing the streams, the familiar Cottonwood and spreading Willows. On the higher plains, and ascending the hills and mountains to their summits, we see a dark forest of lofty Pines; we hear the light breeze sigh and murmur through their branches as it did to the poets of old. But the botanist, in all this array, fails to recognise one solitary acquaintance of his former scenes: he is emphatically in a strange land; a new creation, even of forest trees, is spread around him, and the tall Andes and wide deserts rise as a barrier betwixt him and his distant home.

My indulgent reader will then excuse me, if I, on this occasion, appear before him only as a botanist; culling those objects which have given him so much delight, he wishes to present them to the curious public, alive to the beauties and symmetry of nature's works. Whatever is yet known of their uses and history is also given; and, that the task might be more complete, we have rambled a little beyond, rather than fallen short of, the exact limits of the Republic. We have thus added, as our friends TORREY and GRAY have done, in their general Flora, a collection of the trees of Upper California, extending our ramble as far as the vicinity of Santa Barbara, in about the 34th degree of north latitude. We here met with several Oaks, Pines, a Plane Tree, a Horse-chestnut, and a Box Elder, which have not yet been found within the limits of the Territory of Oregon.

While the work was in progress, Professor TORREY informed me of the arrival of a large collection of dried plants from Key West, in East Florida, made by Doctor BLODGETT, of the United States army. All the trees in this herbarium—at least forty species—were in the most generous manner given up to me for publication by the professor. Most of them form distinguishing features in the tropical landscape of the West India Islands. Among them were the Mahogany, *Simaruba*, the Guaiacum or Lignum-Vitæ, the poisonous Manchineel, several trees of the family of the Myrtles, (*Eugenia*,) three or four species of Fig Trees, the Calabash, and Papaw or Melon Tree, the Mangrove, two species of Cordia, the West India Birch, (*Bursera gummifera*,) and many other arborescent plants which are now for the first time added to the Flora of the United States, and thus in a measure resolving the problem of the geographical limits of the Caribbean Flora. The island of Key West lies about eighty-five miles from East Florida, and is the same distance from Cuba. It is about nine miles long and three broad, containing a population of about four hundred people, chiefly engaged as wreckers.

Besides the trees we have noticed, I have been recently informed of the existence of thickets of *Cactuses* on the island, one of which, with an erect, cylindric, and divided stem, attains the height of thirty or more feet.

In the islands of the Everglades, considerably inland in East Florida, we have been informed that a Palm about ninety feet high, forming a magnificent tree, has been seen; but of this plant we have been unable to obtain, as yet, any further account.

The haste with which I have been obliged to proceed with the publication has prevented me from receiving much advantage from correspondents. Such as have honored me with their remarks are mentioned under the appropriate articles as they occur in the work; and I take this opportunity of tendering them my sincere thanks for all such assistance.

As fast as new materials may be discovered, we intend to give them to the world in the form of a supplement; and we shall then also have an additional opportunity for correcting any errors which may have occurred either in regard to information or in the progress of printing, as well as of making such additions as a more thorough examination of the subject may suggest, particularly the characters of the different kinds of wood indigenous to the most extended limits of the Republic.

Thirty-four years ago, I left England to explore the natural history of the United States. In the ship Halcyon I arrived at the shores of the New World; and, after a boisterous and dangerous passage, our dismasted vessel entered the Capes of the Delaware in the month of April. The beautiful robing of forest scenery, now bursting into vernal life, was exchanged for the monotony of the dreary ocean, and the sad sickness of the sea. As we sailed up the Delaware, my eyes were riveted on the landscape with intense admiration. All was new; and life, like that season, was then full of hope and enthusiasm. The forests, apparently unbroken in their primeval solitude and repose, spread themselves on either hand as we passed placidly along. The extending vista of dark Pines gave an air of deep sadness to the wilderness:—

> "These lonely regions, where, retired
> From little scenes of art, great Nature dwells
> In awful solitude, and naught is seen
> But the wild herds that own no master's stall."

The deer brought to bay, or plunging into the flood from the pursuit of the Indian armed with bow and arrow, alone seemed wanting to realize the savage landscape as it appeared to the first settlers of the country.

Scenes like these have little attraction for ordinary life. But to the naturalist it is far otherwise; privations to him are cheaply purchased if he may but roam over the wild domain of primeval nature, and behold

"Another *Flora* there, of bolder hues
And richer sweets, beyond our garden's pride."

How often have I realized the poet's buoyant hopes amid these solitary rambles through interminable forests! For thousands of miles my chief converse has been in the wilderness with the spontaneous productions of nature; and the study of these objects and their contemplation has been to me a source of constant delight.

This fervid curiosity led me to the banks of the Ohio, through the dark forests and brakes of the Mississippi, to the distant lakes of the northern frontier; through the wilds of Florida; far up the Red River and the Missouri, and through the territory of Arkansas; at last over the

"Vast savannas, where the wandering eye,
Unfix'd, is in a verdant ocean lost;"

And now across the arid plains of the Far West; beyond the steppes of the Rocky Mountains, down the Oregon to the extended shores of the Pacific, across the distant ocean to that famous group, the Sandwich Islands, where Cook at length fell a sacrifice to his temerity. And here for the first time I beheld the beauties of a tropical vegetation; a season that knows no change, but that of a perpetual spring and summer; an elysian land, where nature offers spontaneous food to man. The region of the Bread-fruit; the Tarrow, (*Colocasia esculenta*,) which feeds the indigent mass of the population; the Broussonetia, a kind of Mulberry Tree, whose inner rind, called *tapa*, affords a universal clothing. The low groves produce the Banana, the Ginger, the Turmeric, the inebriating *Kava*, (*Piper methysticum*,) a kind of Arrowroot, resembling the potato, (*Tacca*,) and the Saccharine Tee root, (*Dracæna terminalis*,) at the same time the best of portable fodder. The common timber for constructing houses, boats, various implements, and the best of fuel, is here the produce of a Mimosa, (*Acacia heterophylla*.) For lights and oil, the *too tooc* kernels (*Aleurites triloba*) produce an excellent and inexhaustible supply; the cocoanut and the fragrant *Pandanus* afford deli-

IV.—1*

cious food, cordage, and mats; and the very reeds, reduced in size, which border the rivulets, are no other than the precious sugar-cane of commerce.

Leaving this favored region of perpetual mildness, I now arrived on the shores of California, at Monterey. The early spring (March) had already spread out its varied carpet of flowers; all of them had to me the charm of novelty, and many were adorned with the most brilliant and varied hues. The forest trees were new to my view. A magpie, almost like that of Europe, (but with a yellow bill,) chattered from the branches of an Oak with leaves like those of the Holly, (*Quercus agrifolia*.) A thorny Gooseberry, forming a small tree, appeared clad with pendulous flowers as brilliant as those of a Fuchsia. A new Plane Tree spread its wide arms over the dried rivulets. A Ceanothus, attaining the magnitude of a small tree, loaded with sky-blue withered flowers, lay on the rude wood-pile, consigned to the menial office of affording fuel. Already the cheerful mocking-bird sent forth his varied melody, with rapture imitating the novel notes of his neighboring songsters. The scenery was mountainous and varied, one vast wilderness, neglected and uncultivated; the very cattle appeared as wild as the bison of the prairies, and the prowling wolves, (*Coyotes*,) well fed, were as tame as dogs, and every night yelled familiarly through the village. In this region the Olive and the Vine throve with luxuriance and teemed with fruit; the Prickly Pears (*Cactus*) became small trees, and the rare blooming Aloe (*Agave Americana*) appeared consigned without care to the hedgerow of the garden.

After a perilous passage around Cape Horn, the dreary extremity of South America, amid mountains of ice which opposed our progress in unusual array, we arrived again at the shores of the Atlantic. Once more I hailed those delightful scenes of nature with which I had been so long associated. I rambled again through the shade of the Atlantic forests, or culled some rare productions of Flora in their native wilds. But the "oft-told tale" approaches to its close, and I must now bid a long adieu to the "New World," its sylvan scenes, its mountains, wilds, and plains; and henceforth, in the evening of my career, I return, almost an exile, to the land of my nativity.

CONTENTS OF VOLUME FIRST.

		PAGE
WESTERN OAK	*Quercus Garryana*	14
Holly-Leaved Oak	*Quercus agrifolia*	16
Rocky Mountain Oak	*Quercus undulata*	19
Douglas Oak	*Quercus Douglasii*	20
Dense-Flowered Oak	*Quercus densiflora*	21
Lea's Oak	*Quercus Leana*	25
Dwarf Chestnut	*Castanea alnifolia*	36
Western Birch	*Betula Occidentalis*	40
Oval-Leaved Birch	*Betula rhombifolia*	41
Oregon Alder	*Alnus Oregona*	44
Thin-Leaved Alder	*Alnus tenuifolia*	48
Sea-Side Alder	*Alnus maritima*	50
Opaque-Leaved Elm	*Ulmus opaca*	51
Thomas's Elm	*Ulmus racemosa*	53
Small-Fruited Hickory	*Carya microcarpa*	55
Inodorous Candle Tree	*Myrica inodora*	59
California Buttonwood	*Platanus racemosa*	63
Narrow-Leaved Balsam Poplar	*Populus angustifolia*	68
Long-Leaved Willow	*Salix speciosa*	74
Long-Leaved Bay Willow	*Salix pentandra*	77
Western Yellow Willow	*Salix lutea*	78
Silver-Leaved Willow	*Salix argophylla*	87
Dusky Willow	*Salix melanopsis*	93
California Bay Tree	*Drimophyllum pauciflorum*	102
Large-Leaved Linden	*Tilia heterophylla*	107
American Mangle	*Rhizophora Americana*	112
Florida Guava	*Psidium buxifolium*	115

CONTENTS.

		PAGE
Forked Calyptranthes	*Calyptranthes thytraculia*	117
Small-Leaved Eugenia	*Eugenia dichotoma*	120
Tall Eugenia	*Eugenia procera*	122
Box-Leaved Eugenia	*Eugenia buxifolia*	123
Indian Almond	*Terminalia catappa*	125
Button Tree	*Conocarpus erecta*	128
White Mangrove	*Laguncularia racemosa*	132
Rabbit Berry	*Shepherdia argentea*	134
Mountain Plum	*Ximenia Americana*	138
Osage Orange	*Maclura aurantiaca*	140
Small-Leaved Nettle Tree	*Celtis reticulata*	147
Long-Leaved Nettle Tree	*Celtis longifolia*	148
Cherry Fig Tree	*Ficus pedunculata*	151
Short-Leaved Fig Tree	*Ficus brevifolia*	153
Small-Fruited Fig Tree	*Ficus aurea*	154
Red Thorn	*Cratægus sanguinea*	157
Lance-Leaved Hawthorn	*Cratægus arborescens*	160
Soft-Leaved Cherry	*Cerasus mollis*	164
Holly-Leaved Cherry	*Cerasus ilicifolia*	165
Wild Plum	*Prunus Americana*	169
River Crab Apple	*Pyrus rivularis*	172
American Mountain Ash	*Pyrus Americana*	175
Feather Bush	*Cercocarpus ledifolius*	178
Jamaica Dogwood	*Piscidia erythrina*	180
Broad-Podded Acacia	*Acacia latisiliqua*	183
Blunt-Leaved Inga	*Inga unguis-cati*	186
Guadaloupe Inga	*Inga Guadalupensis*	188
Jamaica Boxwood	*Schæffera buxifolia*	190
Tree Ceanothus	*Ceanothus thyrsiflorus*	193
Snake-Wood	*Colubrina Americana*	195
Carolina Buckthorn	*Rhamnus Carolinianus*	198
Manchineel	*Hippomane mancinella*	202

THE
NORTH AMERICAN
SYLVA.

OAKS.

Natural Order, CUPULIFERÆ. *Linnæan Classification,* MONŒCIA, POLYANDRIA.

QUERCUS. (TOURNEFORT.)

MONŒCIOUS. MALE flowers in loose catkins or racemes. *Calyx* monophyllous, more or less deeply 5-cleft. *Stamens,* five to ten with short filaments, the anthers oval and 2-celled.

FEMALE flower solitary, with a cup-shaped, undivided, hemispherical involucrum formed of agglutinated imbricate scales, sometimes free at the summit. *Perianth* minute, superior. *Ovary* terminated by two to three stigmas, 3-celled, with two ovules. *Nut* or *gland* ovate-cylindric, coriaceous, and smooth, 1-celled; albumen none, germ erect, with thick and fleshy cotyledons.

Trees or shrubs, principally of temperate regions. Leaves alternate, stipulate, simple. Flowers green and inconspicuous, appearing before the complete expansion of the leaves. Nearly allied to the Chestnuts, (*Castanea*.)

WESTERN OAK.

QUERCUS GARRYANA, (Dougl. Mss.) *Foliis petiolatis, obovatis, utrinque obtusis sinuatis subtus pubescentibus, adultis subglabris, lobis obtusis sub-æqualibus superioribus sublilobis, fructibus sessilibus, cupula subhemisphærica dense squamosa, squamis acuminatis pubescentibus, glande ovata.*
QUERCUS *Garryana*, HOOKER, Flor. Bor. Amer., vol. ii. p. 159.

In our western tour across the continent, no feature of the landscape appeared more remarkable, after passing the Mississippi to the wide alluvial borders of the Platte, than the almost total absence of our most characteristic forest trees, the Oaks. When at length we approached the Rocky Mountains, or Northern Andes, we looked in vain for any species of this important genus; and, as far as the eye could trace, we commonly saw nothing but a dark, unbroken mass of gigantic Firs and Pines. It was not till we had nearly reached the shores of the Pacific, that we again beheld any of the familiar features of the Atlantic forest. At the confluence of the Columbia and the Wahlamet we pitched our tents and moored our vessel, which had passed Cape Horn, beneath the spreading shade of majestic Oaks. With the first appearance of extended alluvial plains, immediately below the singular falls of the Oregon, called the Dalles, or Dykes, we observed, for the first time, this Western Oak loaded with its fruit.

The strong resemblance of the leaf of this species to that of the Post Oak (*Quercus stellata*) is almost a libel upon our gigantic plant, which may well rank among the largest of its species. It attains the height of ninety or one hundred feet, if not more, with a diameter of from three to six feet; indeed, amidst a forest the most remarkable in the world for its extreme elevation, our Oak still bore a strict comparison with the

Quercus Garryana.

Western Oak Chêne occidental

rest. Its character, in nearly all respects, equals the famous Oak of Northern Europe, (*Q. pedunculata;*) its lofty summit and enormous branches spread out far and wide, affording the most perfect shade; and, as a picturesque tree, it is much the most striking in the Western landscape. As an object of economy, we found it of the last importance, useful timber trees being scarce along the Oregon; indeed, no other Oak exists along the coast of the Pacific but the present, till we arrive at the absolute boundary of California. How far it extends to the north I am unable to say, but probably as far as Nootka Sound. In Upper California it is scarcely found beyond Monterey; its limit is probably somewhere between the 38th and 50th degree.

The wood is remarkably white for an Oak, hard and fine-grained, and well suited for almost every kind of construction for which the White Oak or English Oak is employed. It was used by our trading party as barrel-staves, and was found no-way inferior to White Oak. Logs of it brought a good price at the Sandwich Islands; and, in short, there is scarcely any thing in which strength or durability are requisite, for which this timber is not suited. The acorns, being sweet and agreeable, form an excellent mast for hogs; and even the aborigines of this region, who never cultivate the soil, employed them for food, first preparing them by stoving and afterward laying them away under ground for future use.

The acorns are much larger than those of the Post Oak, as well as rounder. The leaf bears a considerable resemblance to that species, but is smaller, and, in fact, intermediate in form between it and the European species, (*Q. pedunculata.*) It differs from both in the whiteness of its wood. The bark is whitish and scaly, almost similar to that of the White Oak. The leaves from the first are not pubescent above, or only slightly so along the midrib; the hairs, more numerous beneath, are, as in many other species, collected into stellated clusters;

the young leaves of the Post Oak, previous to expansion, appear brownish-yellow and like a mass of velvet, with the copious pubescence by which they are clad; in ours this appearance never occurs, and the old leaves become nearly smooth; the lobes have narrow, sinuous openings, which scarcely pass halfway down through the leaf; the lobes are usually four on a side, and possess no great inequality with each other; the upper pair mostly present a notch or small division on the lower side, but nothing analogous to the singular obtuse dilatation which that part of the leaf exhibits in the Post Oak. The acorns, besides being larger, are not striated, and the scales of the cup are acuminate, and the upper ones free.

PLATE I.

A young branch, with the leaves not fully expanded, with barren aments.

HOLLY-LEAVED OAK.

QUERCUS AGRIFOLIA. *Foliis lato-ovatis subcordatis dentato-spinosis glabris, fructibus axillaribus sessilibus.*—NEE, in Annal. Scienc. Nat., vol. iii. p. 271. Annals of Botany, No. 4, p. 106.

QUERCUS AGRIFOLIA. *Foliis perennantibus subrotundo-ovatis subcordatis utrinque glabris remote spinoso-dentatis, cupula hemisphærica; squamis adpressis obtusiusculis, glande ovata acuta.*—PURSH, Flor. Bor. Am., ii. p. 657. WILLD. Sp., pl. 4, p. 431.

An Ilex folio agrifolii Americana, forte agria, vel aquifolia glandifera?— PLUKENET, tab. 196, fig. 3.

THIS species, almost the only one which attains the magnitude of a tree in Upper California, is abundantly dispersed over the plain on which Santa Barbara is situated; and, being evergreen, forms a conspicuous and predominant feature in the vege-

Quercus Agrifolia.

tation of this remote and singular part of the Western world. It appears more sparingly around Monterey, and scarcely extends on the north as far as the line of the Oregon Territory. It attains the height of about forty or fifty feet, with a diameter rarely exceeding eighteen inches. The bark is nearly as rough as in the Red Oak; the wood, hard, brittle, and reddish, is used only for the purposes of fuel or the coarse construction of a log cabin.

As an ornamental tree, for the South of Europe or the warmer States of the Union, we may recommend this species. It forms a roundish summit, and spreads but little till it attains a considerable age; as a hedge, it would form a very close shelter, and the leaves, evergreen and nearly as prickly as a holly, would render it almost impervious to most animals.

The leaves vary from roundish ovate to elliptic, and are of a thick, rigid consistence, the serratures quite sharp; the young shoots are covered more or less with stellate hairs, and, for some time, tufts of this kind of down remain on the under side of the midrib of the leaves, which are, however, at length perfectly smooth and of a dark green above, often tinged with brownish-yellow beneath. The staminiferous flowers are very abundant and rather conspicuous, the racemes the length of three or four inches, the flowers with a conspicuous calyx and eight to ten stamens. The female or fruit-bearing flowers are usually in pairs in the axils, or juncture of the leaf with the stem, and sessile, or without stalks. The cup of the acorn is hemispherical, and furnished with loose, brownish scales: the acorn, much longer than the cup, is ovate and pointed.

We do not recollect to have seen this tree properly associated with any other, except, occasionally, the *Platanus racemosus;* their shade is also hostile to almost every kind of undergrowth.

By Persoon, this species is said to have been found on the eastern coast of North America, while Pursh attributes it to the northwest coast, about Nootka Sound. It does not, however, extend even to the territory of Oregon, as far as my observations go. Nee says, "I have only seen branches collected at Monterey and Nootka." The leaves of the young plants (if I am not mistaken) are perfectly smooth when first developed, of a thin consistence, with numerous slender, sharp dentures; beneath they are of a brownish-yellow color, and appear smooth and shining.

PLATE II.

A young branch with barren aments. a. A branch with acorns.

SMALL-LEAVED OAK.

QUERCUS DUMOSA. *Ramis gracilibus pubescentibus; foliis rotundato-ovalibus subsessilibus spinoso-dentatis glabriusculis, subtus villosis concoloribus.*

I OBSERVED this species to form entangled thickets over the base of the hills which flank the village of Santa Barbara, in Upper California. It attains the height of four to six feet, is of a very unsightly appearance, forming what we should call Scrub Oak thickets, of considerable extent, over a barren and rocky soil, which denies sustenance to almost every thing else: the branches divide into many irregular, straggling, and almost naked, slender twigs, clothed with a whitish, smooth bark. The leaves are evergreen, small, and wholly resemble those of the *Quercus coccifera*, but are somewhat pubescent above and softly so beneath; the young twigs are also hairy, with a persisting pubescence. Being unable to discover upon it at the season I visited that country (in the month of April) either flowers or fruit, I am not able to give a figure of it that would be at all interesting.

Quercus Undulata.

Rocky Mountain Oak. Chêne ondulé

ROCKY MOUNTAIN OAK.

QUERCUS UNDULATA. *Fruticosa ramosissima; foliis perennantibus brevipetiolatis oblongis acutis sinuato-dentatis dentibus acutis, basi cuneatis, subtus pulverulento-tomentosis, supra nitidis; fructibus subsolitariis sessilibus, cupula hemisphærica squamis appressis, glande ovata acuta.*— TORREY, in the Annals of the Lyceum of New York, vol. ii. p. 248.

This dwarf Oak, considerably allied to our small-leaved preceding species, was discovered by Dr. James, in Long's Expedition, toward the sources of the Canadian, a branch of the Arkansas, and likewise in the Rocky Mountains. It is said to be a small, straggling shrub, with the under surface of the leaves clothed with a close, whitish tomentum or down, more or less spread, though more thinly, also, on the upper surface, with the hairs stellated. The leaves are small, and somewhat resemble those of the Holly, about an inch and a half or two inches long, rather narrowed at the base, of a thick and rigid consistence, as in all the sempervirent Oaks, reticulately veined beneath, with the margin sinuately toothed, but not, that I can perceive, waved, as the specific name implies; the teeth sharp and acute at the points; above somewhat shining and minutely pubescent. The acorns are large, and strongly resemble those of the Live Oak; they are, however, without stalks, and grow alone or in pairs; the cup is deep and hemispherical, with the scales pointed.

It is so nearly allied to the Holly Oak of the South of Europe (*Quercus Ilex*) that it is necessary to distinguish them. In our plant the base of the leaf is wedge-formed; in the Ilex it is usually rounded, the border less deeply toothed, and not in the least sinuated. The cup and acorn are wholly similar, but in our plant a little larger and less pointed.

PLATE III.

A branch of the natural size, with the acorn.

DOUGLAS OAK.

QUERCUS DOUGLASII. *Foliis membranaceis oblongo-ovalibus basi acutis petiolatis sinuato-pinnatifidis siccitate haud nigrescentibus, supra glabris, subtus puberulis, lobis brevibus acutiusculis, petiolis ramulisque junioribus dense fulvo-pubescentibus; fructibus sessilibus solitariis binisve, cupula hemispherica dense squamosa squamis ovatis convexis in appendicem submembranaceam fulcam appressam linearem obtusam productis pubescentibus; glande ovata cupulam triplo superante obtusa cum umbone conico.*—HOOK. Icon. ined. HOOK. and ARNOT, Bot. Beechy, p. 391.

THIS curious species, of which we have seen only a dried specimen, was collected in Upper California, and bears some affinity to the *Q. Garryana.* According to Hooker and Arnot, the leaves and whole appearance of the plant closely resemble *Q. sessiliflora,* but with different scales to the cup of the acorn. The leaves appear to be smaller, narrower, and less deeply divided than in *Q. Garryana.* The young leaves are covered with down on both sides, and the lobes tipped with short, soft, acute points.

To us, the branch which we have seen bears some resemblance, though vague, to the Post Oak, (*Q. stellata.*) The cup and acorn is also somewhat similar, but larger, while the leaf is smaller and scarcely dilated above. The under surface is

Quercus Douglasii

Douglas's Oak. *Chêne de Douglas*

Quercus Densiflora.

covered with the same stellated pubescence. The stigmas of the fertile flowers are from three to five in number.

PLATE IV.

A branch of the natural size, with acorns. a. The male catkin and young leaf. b. The staminiferous flower magnified.

CASTANOPSIS. Aments elongated and persistent, perianth lanuginous, divided to the base; scales of the spreading cup loose and squarrose; stamens exserted; nut somewhat angular and downy; stigmas several, filiform, and deciduous.

Trees of Oregon, California, and the Himalaya Mountains in India, with the aspect of the Chestnut. Leaves entire, pennately nerved, sempervirent. Aments elongated, erect, the flowers conglomerated. Fertile flowers . . . . ? To this section, or rather genus, belong also, as far as the male specimens are concerned, the *Quercus glomerata* and *Quercus spicata* of Dr. Wallich.

DENSE-FLOWERED OAK

QUERCUS DENSIFLORA. *Foliis perennantibus coriaceis petiolatis oblongo lanceolatis basi obtusis breviter acuminatis parallele nervosis integerrimis margine revolutis junioribus fulvo-furfuraceo-tomentosis subtus pallidioribus demum glabris, amentis masculis elongatis folia superantibus densifloris valde tomentosis nunc ad basin flores paucos femineos gerentibus, fructibus sessilibus, cupula brevi hemisphærica dense squamosa, squamibus elongato-linearibus laxis sericeis, glande ovato-globosa sericea.*—HOOK. Icon. Pl. ined. HOOK. and ARNOT, Bot. Beechy, p. 391.

THIS remarkable tree, scarcely a true Oak, but congeneric with species in the Himalaya Mountains, in India, is a native of

Upper California. It has so much the appearance of a Chestnut, that the cup of the fruit alone attests what it really is. The leaves are evergreen, and of the same lanceolate outline with the Common Chestnut, having similar pennate nerves, but entire, or nearly so, on the margin; at first they are softly clothed beneath with dense, stellate, brownish hairs, but at length become smooth: they are about four inches long and one to one and a quarter wide. The catkins are erect, about four inches long, presenting the appearance of cylindric, woolly spikes, beset with numerous exserted stamens with long, slender filaments, as in the Chestnut. The cup is shallow and patulous, within and without softly sericeous, the scales numerous and acuminate, very loose, somewhat spreading, and two and a half to three lines long. The acorn is large, evidently angular, and more convex on one side, covered with whitish down, and terminated with several filiform, lanuginous, and deciduous stigmas.

The *Castanea chrysophylla* of Douglas, if not the same plant, appears to be another species of this section or genus.

PLATE V.

A branch of the natural size. a. *The acorn.*

OBSERVATIONS ON THE OAKS.

QUERCUS MARITIMA, (*the Maritime Oak.*) The fruit of this species, and sometimes the leaves, approach to the Willow Oak; but this is a low, shrubby plant of the Southern States, with sempervirent leaves, which are very often deeply and distinctly sinuated, rigid, with the lobes often obtuse and mucronate.

OBSERVATIONS ON THE OAKS.

QUERCUS MYRTIFOLIA, (*Myrtle-leaved Oak*.) Of this elegant and curious species, we have yet no materials deserving of a figure.

QUERCUS STELLATA. Q. OBTUSILOBA, MICH. (*Post Oak*.) The variety which I mentioned in the Genera of North American plants, vol. ii. p. 215, under the name of β. *depressa*, rarely exceeds three feet in height, and bears acorns at the height of twelve to eighteen inches from the ground. I first observed it on the hills of the Missouri, up to its confluence with the river Platte, and it is also almost the last species which we find to the westward. I have since met with apparently the same low variety on the gravelly poor hills of the island of Martha's Vineyard, near Massachusetts Bay: it is this scrubby growth of Oak which still affords shelter to the grouse on that island. In some parts of Massachusetts, (according to Emerson,) the usual large growth of this tree is occasionally met with. The species of Quercus which I call *Q. Michauxii* is, I now believe, nothing more than a mere variety of *Q. prinus*.

SWAMP WHITE OAK, (*Quercus bicolor*.) Of this species I first observed a curious variety, which I called β. *mollis* or *Soft-leaved Swamp Oak*, in the swampy elevated forests of the Hudson, near New York; it occurs likewise near Philadelphia and Boston. The leaves, I find, are of the same form as in *Q. bicolor*, but the under side is not white, but partly ferruginous or green, and softly pubescent. The quantity of this clothing, however, varies, and in large leaves it becomes very thin. It forms a somewhat-pyramidal tree, sixty or seventy feet high, branched nearly from the base, the branches deflected and intricately ramified. The leaves are narrowed at the base, and abruptly dilated toward the summit; the dentures are few and sometimes almost wanting; the breadth is about two-thirds of the length; the fruit-stalk or peduncle filiform, two or three inches long.

bearing about one to three acorns on each. It may perhaps be *Quercus filiformis* of Muhlenberg's Catalogue, page 87.

MOSSY-CUP OAK, (*Quercus olivæformis.*) This rare Oak, (which Michaux found only above Albany and in Genesee,) or at least a variety of it with less attenuated cups, is met with in Orange county, New York, where it was observed by Dr. Horton; and it also grows near Vernon, in Sussex county, New Jersey. It has much the aspect of the Water White Oak, (*Q. discolor*,) but the leaves are sinuated.

WHITE OAK, (*Quercus alba.*) According to Emerson, the roots of the White Oak make very beautiful furniture. In England, five pounds sterling have been given for the roots of a White Oak. The pieces have been taken out, and, when sawed and planed, present a wood of extraordinary beauty. A cabinet and table made from the forked branches of this Oak, now in the possession of Mr. C. J. Wister, in Germantown, Pa., may well vie with the finest woods known: it is of a clear, pale yellow, inclining to olive, and feathered in the most beautiful manner; the polish is also equal to that of the finest mahogany.

BARTRAM'S OAK, (*Quercus heterophylla*, MICH. vol. i. pl. 16.) This curious tree, which, in 1837, had attained the height of fifty feet and a circumference of three feet nine inches, was inadvertently cut down, and with it the species, if such it was, appeared to be annihilated; but Thomas G. Lea, Esq., of Cincinnati, informs me "that several years ago he discovered an Oak between two and three miles north of that city, the leaves and fruit of which accord with Michaux's figure. The leaves are sometimes larger than those represented, but with the same outline, irregularly and coarsely toothed, or sub-lobed, and on longish petioles: the margin is very rarely entire. The tree is about twenty-five feet high, and in a vigorous state of growth.

Some scattering Oaks of other species are in its immediate neighborhood. I think it is not a variety of *Q. imbricaria,* many trees of which I have examined, but never found them with leaves the least indented. The *Q. phellos,*" to which it might be allied, "does not grow in the vicinity of Cincinnati, nor, that I know of, in any part of Ohio: this tree, therefore, cannot be a variety of that species." Its nearest affinity appears to me to be to the *Quercus ambigua* of Michaux, Jr., from which it is principally distinguished by the narrower and more simple divisions of its leaves.

LEA'S OAK.

QUERCUS LEANA. *Foliis membranaceis, longissime petiolatis, oblongo-ovalibus, basi rotundatis, subcordatis, sinuato-pinnatifidis, demum glabris, lobis latis integris setaceo-acuminatis; fructibus brevi-pedicellatis; solitariis binisve, cupula hemisphærica, squamis ovatis obtusis, glande subglobosa vittata subsemi-immensa, cum umbone brevi conico.*

OF this remarkably-ambiguous Oak I have already spoken, in a note on *Q. heterophylla,* having at that time, in concert with Mr. Thomas G. Lea, its discoverer, considered it as a variety of that rare species, or some analogous hybrid. Other specimens, accompanied with the ripe glands, have now convinced me that it is either a distinct species or another strange hybrid; but, as I am by no means satisfied of the existence of such spontaneous mixed races among our Oaks, I have taken the liberty of giving it as a species, and dedicating it to its discoverer, an ardent and successful botanist. I shall also take the liberty of adding a quotation from Mr. Lea's notes, made on this plant and sent to me with the specimens.

IV.—2*

"The fruit resembles Michaux's figure of *Q. heterophylla*, but differs in being more depressed and obtuse at the summit. The cups, I think, are alike. The leaves are on longer petioles, but accord in being inclined to be cordate at base. If it is a hybrid, it may have come from the *Q. imbricaria*, or *Q. tinctoria*, or *Q. coccinea*. The fruit is too widely different from *Q. rubra*. The *peduncles* are about the same length as in my specimens of *Q. imbricaria*; in Michaux's figure of that species, the fruit is represented as *sessile*, which I think is wrong. The *petioles* are much longer than in *Q. imbricaria*, the leaves larger and more obtuse at base. These modifications (if it is a hybrid) may be derived from the long petioles and larger leaves of the *Black* and *Scarlet Oaks*. I think it does not partake of *Quercus phellos*, (Willow Oak,) a species that does not grow, to my knowledge, within several hundred miles of this place, (Cincinnati.)

"I saw two individuals of *Q. phellos* in the Bartram garden, which Colonel Carr assured me were propagated from the seed of the original Bartram Oak. Certainly our plant is very like Michaux's figure; but, as that appears to be a hybrid of *Q. phellos*, I think they must be considered distinct. If ours be a hybrid, it most likely comes from *Q. imbricaria* and *Q. tinctoria*, or *coccinea*.

"I have found but a single stock of this, (about five years ago.) It grows three miles north of Cincinnati."

I confess I see too little resemblance in our plant with *Q. imbricaria* to agree with my friend, Mr. T. G. Lea, as to any hybrid connection with that remotely-allied species. Betwixt the Gray Oak (*Q. ambigua*, MICH.) and *Q. tinctoria* I perceive a nearer resemblance. The fruit appears to be wholly that of the Gray Oak. The gland in both is striated, and with a small conic projection. In our plant, however, the base of the gland and that of the cup are yellow, indicating its alliance to *Q. tinctoria*. The leaf differs wholly from both in its simple undivided lobes, though the long petiole and rounded base is that of *tinctoria*. Scarce as this species yet appears to be, under the present circumstances, I am inclined to believe it of a distinct race, with features as distinct as any species in the genus; for the Gray Oak, being, I believe, unknown in Ohio, is again out of the question. I suspect it is in all physical respects allied to *tinctoria*, and would equally afford a yellow dyeing-material.

The full-grown leaves are from five to five and a half inches long by three to three and a half wide, smooth and shining above, with a small quantity of deciduous stellate pubescence beneath. The lobes are about a single pair on a side: the central lobe only sometimes again subdivided into three lesser lobes, all of them ending in bristles. The base is rounded, and often hollowed out, or somewhat sinuated. The buds are small and brown. The fertile flower often by threes, on a short, thick, common pedicle, the middle flower abortive. Male flowers .... not seen. Cups rather deep, as in *Q. tinctoria*, with the scales ovate, obtuse, and closely imbricated. The acorn roundish, somewhat ovate, broadly striate, with a short roundish conic point or umbo about half-way, or nearly so, immersed in the cup.

PLATE V. (*bis*.)

A branch of the natural size with fruit. a. *The cup.* b. *The gland.*

The WILLOW OAK appears to be very nearly allied to the Cluster-leaved Oak of New Spain, (*Quercus confertiflora*,) figured and described by Humboldt and Bonpland; but in that, though otherwise so very similar, the leaves are hairy beneath, while ours are perfectly smooth.

The WILLOW OAK is found as far west as the banks of the Arkansas and several of its branches.

Live Oak, (*Quercus virens.*) Trees near Magnolia, in West Florida, occur of eight to nine feet diameter: it consequently affords large timber. Great quantities of this wood are now brought from the coast of West Florida. According to Wm. Bartram, the Live Oaks on the St. John's in East Florida are from twelve to eighteen feet in circumference; the trunk there rises only from twelve to twenty feet, when it throws out three to five large limbs, which continue to grow in nearly a horizontal direction, each limb forming a gentle curve from its base to its extremity, (*Bartram's Travels*, p. 85;) and he adds, "I have stepped above fifty paces on a line from the trunk of one of these trees to the extremity of the branches." The wood is almost incorruptible, even in the open air. The acorn is small, agreeable to the taste when roasted, and in this state they are eaten by the aborigines as we do chestnuts.

Stately avenues are formed of the Live Oak in South Carolina and Georgia, which, robed in Long Moss, put on an air of sombre grandeur and wildness.

In addition to the geographical limits of the Oaks, I may add that, according to the observations of Emerson, the *Rock Chestnut Oak* (*Quercus montana*, Willd.) occurs in many parts of Massachusetts; he has also found the *Yellow Oak* (*Q. castanea*, Willd.) about Agamenticus Mountain in York, Maine. "It is also found at Saco, in Maine, twenty-five miles farther north." The *Black Oak* (*Q. tinctoria*) "is found in York county, Maine. *Q. palustris* (Pin Oak) is *very* rare in Massachusetts." Mr. Emerson also corroborates my own observations concerning the prevalence of the *Post Oak* on the island of Martha's Vineyard, and adds that "it hardly exceeds twenty inches in diameter and thirty feet in height," which is a circumstance I had overlooked, its prevailing character there being that of a shrub.

The Oaks, though a very extensive genus, are confined to the Northern hemisphere. Besides the numerous species which pervade the United States, sixteen were discovered by Nee in Mexico and New Spain, one of which, the *Q. agrifolia*, is found in Upper California; twenty-one species were added to the Flora of North America by Humboldt and Bonpland, found also in New Spain; four species were discovered in Japan by Thunberg; two in China by Bunge; one in Cochin China, and one in the island of Formosa; two very remarkable species, with lanceolate entire leaves and very long spikes of flowers, like those of a Chestnut, were met with in Nepaul by Wallich; six other species likewise exist in that portion of India; Europe, chiefly the southern part, Northern Africa, and Armenia, afford about twenty-eight species and several varieties; Java, Sumatra, and the Molucca Islands, also produce nineteen species. Thus it appears, of the whole number, according to the enumeration of Willdenow and more recent discoveries, the Old World contains sixty-three species, and North America, including New Spain, about seventy-four. Of these the United States possess about thirty-seven, and New Spain the same number. To these I may also add an additional species from the island of Cuba, nearly allied to our Southern Gray Oak, (*Q. cinerea:*) this I propose to call, after its discoverer, M. La Sagra,

QUERCUS SAGRÆANA. *Foliis perennantibus oblongo-ellipticis obovatisque integris s. sublobatis brevi petiolatis obtusis nitidis margine revolutis subtus tomentosis nervosis subalbidis, fructibus binis pedicellis incrassatis, cupula hemisphærica, squamis appressis, nuce ovata.*

This species apparently forms a tree. The leaves are broader than those of the Gray Oak, of a thick and rigid texture, and are strongly veined both above and beneath; they are about two and one-third inches long and about one inch wide.

Additional Observations. In density and hardness the Live Oak much exceeds every other species of the genus hitherto examined. At first glance, and aided by its great weight, it appears almost like Lignum-Vitæ. The sap-wood is of a pale brownish-yellow, the perfect wood of a pale chestnut-brown, and the extremely fine sawdust almost as bright a brown as that from mahogany. Growing in a climate subject to small changes of temperature, and being evergreen, the woody circles of annual increment are very faint and obscurely marked, which adds to the common density of the fibres. These rings, on young trees, vary from one to two lines in width, but in the older wood they are much narrower. One of the most striking features of this wood, however, is the distinctness of the medullary rays, which traverse in strong and pale lines the faint waves of the annual increments. For the first forty or fifty years, the Live Oak appears to increase in the bulk of its trunk as fast as our White Oak; but after that period the growth is much more slow; still, the density of its wood is so great, that, through a strong magnifier, the pores and vessels are barely visible. In the United States Navy Yard, in this place, I have measured a squared log of Live Oak, thirty-two feet long, which probably formed the trunk of a tree not less than fifty to sixty feet in height. The present value of moulded Live Oak varies from $1.20 to $1.30 and $1.45 per cubic foot. Promiscuous unprepared logs sell from $1.20 to 98 cents and $1 the cubic foot. Some very choice timber sells as high as $1.65. This valuable timber has been employed in the United States navy between fifty and sixty years.

Little is yet known respecting the southern limits of this species of Oak, though there can be little doubt that it continues along the borders of the Mexican Gulf to Yucatan. Dr. Burroughs informs me that it is said to be found growing on the banks of the Alvarado River, about seventy-five miles south of Vera Cruz. I am also informed of the existence of the Live Oak near Matagorda in Texas.

It is stated in a late Texian paper that an English company have recently landed on the Brazos, in the neighborhood of

ADDITIONAL OBSERVATIONS.

Brazoria, for the purpose of getting out Live Oak. They are said to have contracted with the English Government to deliver two millions of cubic feet. The country about Brazoria is loaded with enormous trees, some of them casting a shade of one hundred and fifty feet in diameter. The Live Oak extends into Texas at least one hundred and fifty miles, according to the observations of Dr. Casper Wister, Jr., of Germantown, Pa.

John Lenthall, Esq., United States Naval Constructor, has favored me with the following remarks concerning the timber used in the United States navy.

The frames and principal pieces are all of Live Oak; and the frames of several of our ships that were cut from the islands of Georgia and on the coast, thirty years since, are still in an excellent condition, though in some ships, in which the timber was cut inland, the result is not so favorable. The weight of a cubic foot varies from seventy-three to seventy-eight pounds. This timber is peculiarly adapted to ship-building, and is scarcely fit for any thing else, being short and crooked, so that the timbers are rarely grain-cut.

The White Oak, used almost exclusively for plank, is cut from the seaboard of the Middle States, and is equal to the best English or foreign timber. The Red Oak is never used. The Oak from Canada is that which has generally been introduced into England, and from it a very erroneous opinion has been formed with regard to the Oak timber of the United States, for the Northern timber is much inferior to that from the Southern States, and is never used. A cubic foot of unseasoned White Oak weighs from fifty-eight to sixty pounds, and when seasoned, forty-seven to forty-nine pounds. White Oak timber is often brought from the Lakes and used for keels and bottom-planks; but for upper works that from the Delaware and Chesapeake Bay is preferred, being much stronger and more durable. This Lake timber is principally to be found at New York.

From the Delaware River and Chesapeake Bay large quantities of White Oak are likewise shipped for the Eastern States, of which the better class of ships are built. A great deal of Pine timber is also shipped from thence for the same purpose.

The Quercitron is the bark of the *Quercus tinctoria*, freed from the epidermis. Besides tannin, it contains a yellow coloring-matter, which may be extracted by water, and which, on evaporation, yields a peculiar extract to the amount of eight per cent. of the bark employed. The tannin belongs to that variety which precipitates iron of a green color. This tannin is very injurious to the color, because it is precipitated by the same reagents with the color, and imparts to it a brownish tint. To obtain the coloring-matter free from it, a bladder softened in water, and cut into small pieces, freed from all the parts which are soluble in water, is applied to the infusion of the Quercitron bark, which takes up the tannin; or it may be precipitated by a solution of isinglass.

According to Chevreul, the coloring-matter which he calls quercitrin, although not a simple substance, is obtained by cautiously concentrating an infusion of Quercitron. A crystalline substance then precipitates, which, while yet in suspension in the liquid, imparts to it a pearly appearance. It exhibits a slight acid reaction by curcuma-paper. It is slightly soluble in ether, but more completely so in alcohol. Water dissolves it; and the solution becomes orange-yellow by the addition of alkali. The acetate of lead and of copper, as well as the protochloride of tin, precipitate it in yellow flakes. Sulphate of the peroxide of iron colors it at first olive-green, and then causes a precipitate. Sulphuric acid dissolves quercitrin, and the greenish-orange colored solution becomes cloudy by the addition of water. By dry distillation it yields, among other products, a liquid which soon crystallizes, the crystals possessing all the properties of quercitrin.

In the dyeing-establishments the clear yellow color is obtained by precipitating the tannin by means of a solution of glue or buttermilk; the coloring-matter then remaining in the solution is mixed with the solution of alum and carbonate of potash, by which it is precipitated of a yellow color in combination with the alumina. Protochloride of tin also produces with it a strong yellow precipitate.

QUERCUS RUBRA? The largest Red Oak in North America, says a correspondent of the Natchitoches Herald, can be seen on the plantation of W. Smith, Esq., eighteen miles from Natchitoches, on the road leading to Opelousas. This majestic Oak stands in the midst of a rich and heavy bottom, on the Bayou St. Barb. Two feet from the ground it measures *forty-four* feet in circumference, and at six feet, *thirty-two* feet. The trunk appears sound and healthy, and its height, *to the branches*, is from fifty to sixty feet.

From Dr. G. Engelmann, of St. Louis, I learn that the White Oak (*Q. alba*) and the Rock Chestnut Oak (*Q. montana*, WILLD.) grow in that vicinity, where there are two varieties of each with sessile and with pedunculated fruit, in this respect agreeing with the *two* varieties of the English Oak, (*Q. robur*,) which have been considered as two species. He also informs me that the Chinquepin Oak (*Q. prinoides*, WILLD.; *Q. prinus chincapin*, MICH. Sylva, t. 11) grows commonly in Southwestern Missouri. He also adds, that the Spanish Oak (*Q. falcata*) he has only seen in the southern extremity of Missouri; and that the Water Oak (*Q. aquatica*) grows no nearer to him than the banks of the Arkansas.

The Sweet Gum Tree (*Liquidambar styraciflua*) he saw on the borders of the Wabash; it grows also in Southern Missouri, and all through Arkansas to the province of Texas; but he has not seen it through the greater part of Missouri and Illinois.

The Black Gum Tree, (*Nyssa multiflora*, Walt. N., *Sylvatica*,

Mich. Sylva, t. 110,) according to Dr. Engelmann, is common in the southern parts of Missouri.

The Wahoo Elm (*Ulmus alata*) Dr. Engelmann finds as far north in Missouri as the vicinity of Herculaneum.

Around Cape Girardeau, one hundred and fifty miles south of St. Louis, he also observes the Tulip Tree, (*Liriodendron*.)

Beech Trees, the doctor informs me, he has not seen west of the Wabash, except near Cape Girardeau: they grow associated with Pines in Western Louisiana, and I have seen them in the forests which border the Arkansas.

CHESTNUT.

Natural Order, AMENTACEÆ, (Juss.)   *Linnæan Classification,*
MONŒCIA, POLYANDRIA.

CASTANEA.* (TOURNEFORT.)

POLYGAMUS. The *male* ament elongated, composed of numerous interrupted clusters of flowers, with a five or six-parted perianth. Stamens ten to twenty. *Female* flowers about three in an ovoid muricate valvular involucrum. Perianth urceolate, 5 or 6-cleft, having rudiments of abortive stamens. The ovary incorporated with the perianth, the stigma pencillate, exserted, its divisions rigid and pungent. Nuts one to three, included in the enlarging echinate, 4-cleft involucrum.

These are trees or shrubs of temperate Europe and North America, with alternate, stipulate, mucronately-serrated leaves, and very long, axillary aments. Nuts farinaceous, edible.

---

* So named from *Castanea*, a town of Thessaly, near the river Peneus, where large Chestnut Trees are still found.

DWARF CHESTNUT.

Castanea alnifolia. *Depressa, foliis obovatis subacutis, mucronato-serratis subciliatis junioribus subtus pubescentibus, amentis filiformibus solitariis tomentosis.*
β Pubescens. *Foliis brevioribus, adultis subtus pubescens.*
Castanea alnifolia, Nutt., Gen. Am., vol. ii. p. 217.
Castanea nana, Elliott, Sk., vol. ii. p. 615, (not of Muhl.)
Fagus pumila, var. *præcox*, Walter, Carolin., p. 233.

A species remarkable for its dwarf growth, and inserted only to complete the history of the genus. It rarely exceeds a foot in height, growing in small patches, with creeping roots. I first met with the variety β in the vicinity of Charleston, South Carolina; afterward the smoother kind, much more abundant, and in flower in the month of March, round Tallahassee, in West Florida.

The Floridian plant is scarcely a foot in height, with smooth, purplish-gray branchlets; the leaves obovate, on very short petioles, deeply serrate, obtuse or acute, elliptic-obovate; when young, whitish pubescent; the adult almost perfectly smooth on both surfaces; about three inches long by one inch or more wide. Stipules subulate, rather persistent. Male aments solitary, long, and filiform, tomentose. The fruit I have not seen.

The Charleston plant grows in sandy pine-barrens, and the nut, which is solitary, is said by Elliott to be much larger, but less abundant, than in the other native species. This plant rarely exceeds two feet in height. Its leaves are glossy above, pubescent, but not tomentose, beneath. Fertile flowers one to three in an involucrum, only one perfected.

The wood of the Chinquepin, (*C. pumila*,) whenever it can be obtained large enough for posts, is much valued, as it is sup-

Castanea alnifolia.

Dwarf Chestnut. Châtaignier à feuilles d'aune.

posed to be more durable when exposed to the weather than any other timber except the Red Cedar.—ELLIOTT.

PLATE VI.

A branch of the natural size.

GOLDEN-LEAVED CHESTNUT.

CASTANEA CHRYSOPHYLLA, (Dougl. Mss.) *Foliis sempervirentibus lato-lanceolatis acuminatis coriaceis integerrimis glabris subtus aureo-farinosis.* HOOK., Flor. Bor. Am., vol. ii. p. 159.

ACCORDING to Douglas, this is a splendid evergreen tree, varying in height from twenty to seventy feet, with leaves four to five inches long, deep green above, and below of a rich golden yellow. These leaves are, also, (very different from all the rest of the genus) quite entire. The spikes or catkins of the flowers scarcely exceed an inch in length, including the peduncle, and they are solitary in the axils of the upper leaves. Sometimes all the flowers on a catkin are male; sometimes the two or three lower flowers are female. The fruits are two or three, crowded or densely covered with acicular prickles. Said to be common at the Grand Rapids of the Columbia, Cape Orford, and near Mount Hood; constantly affecting the hills. This species rests wholly on the authority of Douglas. I did not meet with it, nor does it appear that any specimens were sent to England. It will probably prove to be some very different genus to that of the present.

Additional Observations. In regard to the Western range of our forest trees, Dr. Engelmann informs me, by letter, that, though the Chestnut (*Castanea Americana*) does not grow in the immediate valley of the Mississippi, it still reappears again in Southwestern Missouri and the northwestern portion of Arkansas, where is also found the Locust Tree, (*Robinia pseudacacia.*)

CHESTNUT TREE, (*Castanea Americana.*) The wood of this tree is capable of receiving a fine polish, and well-selected pieces present waves and feathered figures of considerable beauty and variety, the more striking as they are seen with great distinctness through a pale and light ground. Furniture of this kind may be seen at Mr. Crout's cabinet-warehouse in Philadelphia.

BIRCH.

Natural Order, BETULINEÆ, (Richard.) *Linnæan Classification,* MONŒCIA, POLYANDRIA.

BETULA.* (LINN.)

Male flowers in long, cylindric aments. Scales in a double series, the inner by threes, 1-flowered; stamens six to twelve. *Female* flowers with ovoid or oval aments; the scales trifid, 1 to 3-flowered. *Styles* two. *Nuts* minute, compressed, 1-seeded, edged with an alated, thin margin.

Trees or shrubs of the colder parts of the northern hemisphere on both continents, with the bark often exfoliating in thin, circular plates. Leaves alternate, ovate, or deltoid, serrated; producing stipules; aments axillary.

---

* Supposed to be derived from *Betu,* the Celtic name for the Birch.

WESTERN BIRCH.

BETULA OCCIDENTALIS. *Ramis resinoso-verrucosis, foliis lato rhombeo-ovatis sublobatis inciso-serratis hirsutulis, subtus pallidioribus punctatis, nervis rematis, amentis femineis lato-cylindraceis squamis lobis lateralibus ovatis intermedio longiore.*
BETULA *occidentalis*. HOOK., Flor. Bor. Am., vol. ii. p. 155.

This low species of Birch, only six to ten feet high, was first observed westward near the sources of the Sweet Water, a northern branch of the Platte, and where it penetrates into the first range of the Rocky Mountains. On the borders of this clear stream, diminished to a small, purling brook, and accompanied by clumps of willows, we first saw it growing. According to Drummond, it occurs on the east side of the Rocky Mountains down to Edmonton House. Douglas found it near springs on the west side of the Rocky Mountains; and Dr. Scouler met with it in Oregon, near to the Straits of Juan de Fuca; it also grows near Walla-Walla, and continues up the Oregon to the country of the Flatheads.

The principal branches are erect and somewhat virgate, clothed with a bright brown bark, copiously sprinkled with small resinous warts, so as to render the branches rough to the touch. The leaves are somewhat deltoid, or rhomboidly-ovate, on shortish petioles, in my specimens acute, but not acuminate, sharply and somewhat unequally serrated, and very slightly lobed; above, somewhat glutinous, with very few pinnated nerves; below, paler; the midrib and nerves sprinkled with a few long hirsute hairs, which are also seen above, on, and near the petiole. The leaves, in flowering specimens, are only about one and a half inches long by an inch wide. (The adult leaves described by Hooker are much larger, two to two and a

Betula occidentalis.

Western Birch.   Bouleau occidental

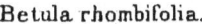

Betula rhombifolia.
Oval leaved Birch.   Bouleau à feuilles ovales.

half inches long.) The aments are cylindric, in the staminiferous plant, composed of a double series of scales. Female aments pedunculated, cylindric, at length drooping, often accompanied by a very small leaf at the base; the scales trifid and dilated, strongly ciliated, the lateral lobes ovate; the central one nearly linear and longer; three germs beneath each scale. Nuts broadly winged. Styles two, very long and subulate; summit of the germ pubescent.

The trunk of this species is only a few inches in diameter, so that it scarcely ranks with proper trees. The leaves are bitter to the taste.

PLATE VII.

A branch of the natural size. a. The seed-vessel.

OVAL-LEAVED BIRCH.

BETULA RHOMBIFOLIA. *Ramis resinosis gracilibus, foliis subrhomboideoovalibus, vix acutis, grosse serratis, subtus pallidioribus epunctatis; venis pilosiusculis amentis femineis cylindraceis, squamis tripartitis glabriusculis lobis ovatis, lateralibus brevibus.*

THIS is a still more humble shrub than the preceding, which it somewhat resembles. It grows in the central Rocky Mountain range, and continues more or less to the banks of the Oregon. It is spreading and somewhat decumbent, with slender brown twigs, which, when young, are more or less covered with resinous atoms. The leaves, with their petioles, which are two or three lines, are not more than an inch long by half an inch wide, oval, and somewhat rhombic, deeply, sharply, and almost equally serrate, rounded, but still generally acute, smooth above, paler beneath, with a very few distant nerves, somewhat hairy along their margins beneath. External scales of the male aments ovate and ciliate. Stamens about six. Female aments with nearly smooth, deeply 3-parted scales, of which the central division is the longest. I have not seen the ripe fruit.

PLATE VIII.

A branch of the natural size. a. The seed-vessel.

Observations. On the summit of the White Mountains of New Hampshire grows the *Betula nana* of Europe, found there by Mr. Oakes as well as myself.

Dr. Charles Pickering also collected a specimen on those mountains, which appears to be the *Betula fruticosa* of Pallas, first found in Siberia.

In the Rocky Mountains, besides the two species now described, we met with the *Betula glandulosa*, which is also found on the high mountains of the United States.

CANOE BIRCH, or PAPER BIRCH, (*Betula papyracea.*) This very useful species of Birch to the aborigines of the North is found, according to the observations of Dr. Richardson, as far as the 65th degree of latitude.

WHITE BIRCH, (*Betula populifolia.*) Hooker very justly remarks the near affinity which this species bears to the common European Birch, (*B. alba.*) The general aspect is the same. In our plant, however, the leaves have longer acuminated points and smaller fertile catkins. The scale of the same catkin in ours is also comparatively smaller and shorter-clawed, with the middle lobe acute and much smaller than the lateral lobes; whereas, in the European Birch, the lobes are nearly all equal and obtuse.

ALDERS.

Natural Order, AMENTACEÆ, (Juss.) *Linnæan Classification,* MONŒCIA, TETRANDRIA.

GENUS ALNUS. (TOURNEFORT, DECANDOLLE.)

CHARACTER. The flowers are *monœcious,* (or of two different kinds on the same plant,) disposed in catkins, (or cylindric spikes of short duration;) those producing the stamens are long and cylindric; those of the fruit or seed are ovoid or globular, produced upon branching peduncles. The scales of the *male* flower are pedicellated, and in the form of an inverted heart, bearing beneath each three lesser scales; the proper flowers are situated at the base of each of these, and are composed of a cup with four lobes and four stamens. The scales of the fruiting catkins are wedge-shaped, hard, and persistent. The ovary is compressed, and bears two long stigmas. The envelop of the seed is hard, with a border which is either thick or membranaceous, and presents two cells with two seeds: the ovules in the germ are about four, or two in a cell, three of them usually abortive.

The plants of this small genus, confined to the temperate or colder parts of Europe and North America, are either shrubs or trees, with deciduous leaves, generally growing by streams, or in cool and humid places. As trees, they seldom attain a greater elevation than thirty to forty feet; the wood is hard and yellowish, becoming of a brownish red, nearly like mahogany, when exposed to the air, and capable of acquiring a fine polish. When stained black, it resembles ebony; and it is capable of enduring moisture for a great length of time.

The Alders may be divided into the two following sections: in both the peduncles are subdivided.

§ I. *The seed-vessel furnished with a membranaceous winged margin, and with the scales of the fertile ament retuse or obscurely lobed.*

    The White Alder, (*Alnus incana*.)
    The Oregon Alder, (*Alnus Oregona*.)
    The Heart-leaved Alder, (*Alnus cordata*.)
    Mountain Alder, (*Alnus viridis*.)

§ II. *The margin of the seed-vessel thick and opaque, and with the scales of the fruiting ament distinctly lobed.*

    Common Alder, (*Alnus glutinosa*.)
    Fine-toothed Alder, (*Alnus serrulata*.)
    Sea-side Alder, (*Alnus maritima*.)
    Oblong-leaved Alder, (*Alnus oblongata*.)
    Short-leaved Alder, (*Alnus brevifolia*.)
    Rhombic-leaved Alder, (*Alnus rhombifolia*.)

---

§ I. *Fruit alated.*

OREGON ALDER.

ALNUS OREGONA. *Foliis lato-ovatis utrinque acutis, duplicato-serratis junioribus glutinosis, venis subtus pubescentibus pallidis; stipulis oblongis deciduis glutinosis, ramulis glabris.*

ALNUS *glutinosa.* PURSH, Flor. Bor. Am., vol. ii. p. 623, (in part.)

THIS tree, like the Common Alder of Europe, attains the height of thirty or forty feet, with an erect, smooth trunk of small diameter, the wood of which is very similar to that of the European species, and might, no doubt, be employed for the same purposes: it is, however, lighter in color, but of a close grain.

Alnus Oregona.

As an ornamental tree, it is well worth attention, producing an elegant erect top, and affording considerable shade by the largeness of its leaves, which are about three inches long by two and a half wide. We found it, as usual with the plants of this genus, growing along the borders of small, clear brooks, near the confluence of the Wahlamet, but seldom, if ever, on the banks of the larger streams which are subject to inundation. In our progress to the West, we first observed this tree on the borders of the rivers Boisée and Brulée, which pass into the Shoshonee not far from Walla-Walla, and at intervals it continues more or less common to Point Chinhook, near the shores of the Pacific.

The twigs are smooth and of a brown color, and the young buds of every kind resinous, as well as the upper surface of the younger leaves; beneath, the leaves are more or less pubescent, particularly along the veins, and paler and often somewhat ferruginous. The veins are very strongly marked and prominent beneath, the teeth large, but the denticulations minute and glandular at the points. Sometimes the leaves are elliptic-ovate, the stalks about the third of an inch long. The stipules are resinous, and disappear with the evolution of the bud. The fruiting ament is roundish-ovoid, and very similar to that of the Common Alder.

This species is nearly allied to the White Alder (*Alnus incana*,) but differs sufficiently in its buds, branchlets, stipules, and leaves; in both the fruit is provided, as in the Birch, with a translucent, membranous wing. It appears, likewise, to have a considerable affinity to *A. acuminata* of Humboldt and Bonpland, a tree of Peru, discovered by Dombey, but in that species the leaves are more lanceolate than ovate and acuminate.

Besides the other economical uses for which the wood of the Alder is employed, the knots furnish a beautifully-veined wood for cabinets; handsome chairs have been made of it, which acquire the color of mahogany. In France it is used in making sabots, or wooden shoes, and in the North of England it is employed for the thick soles of a kind of shoes called clogs, and is preferred for these uses, in consideration of its durability and lightness. The chips, boiled with copperas, give a black dye to wool, and the leaves have been used in tanning; sheep will browse on them and on the smaller branches.

PLATE IX.

A branch of the natural size. a. The seed-vessel.

WHITE ALDER.

ALNUS INCANA. *Foliis oblongis acutis subtus pubescentibus, axillis venarum nudis, stipulis lanceolatis.* WILLD. Sp. pl.
Alnus undulata, WILLD., Sp. pl., vol. iv. p. 336.
Black Alder, (*Alnus glauca,*) MICH., Sylva, vol. i. p. 378.
Betula Alnus, crispa, MICH., Flor. Bor. Am., vol. ii. p. 181.
Betula crispa, AITON, Kew., vol. iii. p. 339.
Betula alnus, β. LINN., Sp. pl

THIS species forms a much smaller tree than the Common Alder, being only twelve to eighteen feet high, and sometimes indeed a mere shrub, as in the Alleghany Mountains in Pennsylvania. In Massachusetts and Maine it attains its greatest size. Its bark is gray or cinereous: the leaves are sometimes villous beneath, and the stipules persistent after the development of the leaves, which are noway glutinous; those of the young plants are smooth and glaucous beneath. It is common to the mountainous parts of Europe not less than to the northern parts of the United States. It occurs likewise in this vicinity.

MOUNTAIN ALDER.

ALNUS VIRIDIS, (DECANDOLLE.) *Foliis rotundato-ovatis irregulariter argute serratis glabriusculis, stipulis ovatis membranaceis deciduis; fructibus late alatis.*
Alnus viridis, DECANDOLLE, Flore Française, vol, iii. p. 304.
Betula viridis, VILLARS, Dauphin, vol. iv. p. 789.
Betula ovata, SCHRANK, Salisb., p. 25.
Betula incana, β. LAMARCK, Dict., vol. i. p. 455.
Alnus Alpina minor, BAUHIN'S Pinax, p. 428.

LABRADOR, the elevated summits of the White Mountains of New Hampshire,[*] and the tops of the high mountains of North Carolina,[†] are the only localities on this continent where the Mountain Alder has yet been found. It occurs likewise in the Alps of Switzerland, at an elevation of between four and five thousand feet above the level of the sea, where it frequently forms a small tree about six feet in height. In the White Mountains its stature is much more depressed, and it is of rare occurrence. It grows likewise in the barren and cold climate of Kamtschatka.

The wood is white, and the branches are covered with a cinereous smooth bark. The leaf is near two inches long and one and a half wide, nearly smooth on both sides, but generally somewhat hairy along the veins beneath, rather acute, with numerous sharp, small, and irregular serratures, but not doubly serrate. The male catkins are long, and grow, two or three

---

[*] A specimen in the herbarium of the Academy of Natural Sciences in Philadelphia was discovered on the White Mountains by my friend, Dr. Charles Pickering, at an elevation of four thousand feet above the level of the sea.
[†] Recently discovered by Dr. Gray and Mr. Curtis.

together, at the extremities of the twigs; each scale contains three tetrandrous flowers, as usual in the genus. The fertile aments are roundish and elliptic, about three together, and terminal; the scales are truncated and obscurely lobed at the extremities; the fruit, like that of the Birch, is furnished with a broad, thin, conspicuous winged margin.

§ II. *Fruit not alated; the margin opaque.*

THIN-LEAVED ALDER.

ALNUS TENUIFOLIA. *Foliis lato-ovatis subacutis duplicato-crenatis glabris basi rotundatis longe petiolatis, stipulis deciduis, pedunculis femineis duplicato-ramosis.*

This very distinct species of Alder, which arranges with our common species, (*A. serrulata*,) was met with on the borders of small streams within the range of the Rocky Mountains, and afterward in the valleys of the Blue Mountains of Oregon, a chain which may be called, as it were, in comparison of their elevation, the Alleghanies of the West.

This species falls short of the character of a tree; but yet it is scarcely inferior in size with our common species, growing to about the height of a man, with numerous short branches covered with a smooth gray bark. The leaves are about two inches long by one and a half wide, with slender petioles, from a half to three-quarters of an inch in length; they are of a thin consistence, and usually smooth, with obtuse denticulations. The fruiting-branches are often subdivided, each branch bearing from three to five small, roundish, ovate aments, of which the

Alnus tenuifolia.

Thin leaved Alder. Aune menu feuillé

scales are very distinctly lobed. The fruit is unusually small and elliptic, terminated by the two remaining styles, and having a thin, opaque margin.

PLATE X.

A branch of the natural size. a. The fruit.

RHOMBIC-LEAVED ALDER.

ALNUS RHOMBIFOLIA. *Foliis subrhomboideo-ovatis obtusiusculis glutinosis basi acutis, subduplicato-serrulatis serraturis crebris acutis, subtus puberulis axillis venarum nudis, stipulis oblongis membranaceis deciduis.*

I OBSERVED this species, a large shrub, in the vicinity of Monterey, in Upper California. Its nearest relation appears to be to the European Alder, (*A. glutinosa,*) from which, however, it is abundantly distinct. The fruit I have not seen.

The leaves are about two inches long and one and a half wide, glutinous, beneath nearly the same color as above, and pubescent along the veins; the petioles are not more than about two lines long. The twigs are smooth and brownish. It appears to be allied to the oblong-leaved Alder (*A. oblongata*) of the South of Europe.

SEA-SIDE ALDER.

ALNUS MARITIMA. *Foliis ovalibus glabris serratis obtusis vel acuminatis, basi acutis, subtus ferrugineis; amentis femineis maximis, squamis duplicato-lobatis.*

ALNUS MARITIMA. *Foliis ovatis serratis, basi acutis.*—MUHL., MSS. Observationes Botanicæ de Plantis Am. Septent., p. 193, (in the Library of the Acad. Nat. Sc. Philad.) and Herbarium.

A SPECIMEN of this very distinct species of Alder was collected on the Eastern Shore of Maryland, by my friend, Charles Pickering. It has the appearance of being a low shrub, with slender, smooth branches. The leaves are two and a half to three inches long by one and a half or more wide, of an elegant, well-defined, oval outline, and supported upon longish petioles; the young buds and leaves, after the manner of the genus, are slightly glutinous; the nervings very slender, serratures shallow, and in the larger leaves rather remote; the uppermost leaves on the infertile shoots are acuminated, and, at first glance, look almost like the leaves of a Camellia. The male catkins are unknown, as are the stipules, which are probably small. The fertile ament, in size and general appearance, might be taken for the strobile or cone of a Spruce; it is about the size of a Hop-cluster, nearly black, with the scales very thick and deeply and obviously lobed. The carpel is small in proportion, and with a thick, opaque, and obscure margin, as in *A. serrulata*.

PLATE X.

A branch of the natural size. a. The seed-vessel.

Alnus maritima.

Sea side Alder　　　　　Aune maritime

Ulmus opaca.

Opaque leaved Elm.   Orme opaque.

ELMS.

Natural Order, ULMACEÆ, (Mirbel.) *Linnæan Classification*,
PENTANDRIA, DIGYNIA.

ULMUS. (LINN.)

OPAQUE-LEAVED ELM.

ULMUS OPACA. *Foliis parvis oblongo-ovatis obtusis scabris, subduplicato denticulatis, basi cuneatis obliquis subtus pubescentibus, floribus fasciculatis, fructibus hirsutis.*

IN the summer of 1818, on my journey into the interior of the territory of Arkansas and on the plains of Red River, near its confluence with the Kiamesha, 1100 miles up the former stream, I had the satisfaction of discovering this curious Elm, which, like our other species, forms a majestic and spreading forest tree of the dimensions of an ordinary Oak. In those dry and open savannas, the shade of this densely-verdant tree proved more than usually acceptable. It is remarkable for the smallness and thickness of its oblique and usually blunt leaves, which, with their short stalks, are only about an inch in length by half that dimension in breadth; they are also very numerous, close together, scabrous, with minute papillæ, of a deep green above, and somewhat shining, oblong-ovate, mostly obtuse, the margin with shallow, double denticulations; beneath, the leaf is paler, a little brownish, with strong pennate, simple, or forked nerves; the base of the leaf is oblique, as well as the whole outline, and one half of the leaf is much narrower than the other; the nerves are pubescent. The young branches are smooth and brownish. The leaves, before complete development, are canescently tomentose and attended by large, oblong, membranous, brown stipules. The taste of the plant is astringent, but noway mucilaginous.

This remarkable species appears to be nearly allied to *Ulmus chinensis*, judging from the short description in Persoon and Duhamel. The flowers are fasciculated in small numbers and on short peduncles. The samara is elliptic, rather deeply bifid at the summit, covered with a dense and somewhat ferruginous pubescence even when ripe.

Of the uses and quality of the timber of this species I am unable to speak from experience, as it grew remote from the settlements at that time established in the territory. The density of shade produced by it, so crowded with rigid leaves, and the peculiarity of its appearance, entitle it to a place in the nurseries of the curious, and it is probably quite hardy enough for all temperate climates. To this species Virgil's epithet—

"Fœcundæ frondibus ulmi"—

might more justly be applied than to any other.

PLATE XI.

A branch of the natural size.

Elmus racemosa.
Orne à grappe

THOMAS'S ELM.

ULMUS RACEMOSA, (THOMAS.) *Foliis ovatis acuminatis duplicato-serratis glabris subtus pubescentibus; floribus racemosis fasciculatis.*

ULMUS RACEMOSA, flowers in racemes; pedicles in distinct fascicles, united at their bases; leaves ovate, acuminate, doubly serrate, glabrous above, [minutely] pubescent beneath; stigmas recurved. EATON'S North Am. Bot., (ed. 8.,) p. 464. THOMAS, in Silliman's Journ. Sci., vol. xix. p. 170, with a Plate.

THIS species, confounded with our other Elms, is, according to Professor Torrey, an abundant species in the western part of the State of New York, and, probably, of the Western States generally. Mr. Thomas, its discoverer, found it in Cayuga county, in the State of New York, and in the adjacent country. According to Emerson, he believes Mr. Oakes has obtained specimens from Vermont, collected by Dr. Robbins, so that it is probably a Northern and Western species.

The lower, stout branches, according to Mr. Thomas, produce corky excrescences like the Wahoo Elm. Leaves broad-ovate, acuminate, obliquely auriculated on one side, doubly serrate, smooth, and somewhat shining above, with the under surface and ribs minutely pubescent. The flowers, unlike any other Elm, are disposed in racemes, composed of several clusters of two to four together, and extending to the length of one to two and a half inches, often furnished with one or two small but perfect leaves before the terminal buds are open; the flowers distinctly pedicellate. Calyx 7 to 8-cleft. Stamens seven to ten. Stigmas two, recurved. Samara elliptic, large, and very pubescent, with the margin thickly fringed, and the membrane more extended on one side as indicative of a second but abortive cell.

PLATE XII.

A branch of the natural size. a. The flower. b. A branch with the corky bark.

HICKORY.

Natural Order, JUGLANDEÆ, (Decand.) *Linnæan Classification*, MONŒCIA, POLYANDRIA.

CARYA.* (NUTT., Gen. Am.)

Staminate flowers in very long and loose, ternate aments, scales imbricated, 3-parted. Stamens three to six, with pilose anthers. *Fertile* flower with a single 4-cleft superior, herbaceous perianth. Style none; stigma partly discoid, 2-lobed, the segments bifid. Pericarp woody, 4-valved. Nut mostly somewhat quadrangular, with an even surface.

Large trees of North America, confined to the eastern side of the Rocky Mountains, and extending from Upper Canada to Florida. Leaves alternate, unequally pinnate, without stipules. Flowers polygamous, in compound, pendulous, pedunculated aments appearing with the leaves; female flowers terminal; the pericarp opening by four valves. Nuts edible or bitter, usually more or less quadrangular; in the *Pakan*, even. Pubescence tufted or stellate. Persoon, as far back as 1807, divided the genus Juglans into the two natural sections which it presented: his second division included the Hickories only. " * * *Amentis masculis compositis, tetrandris.*"

* From καρυα, the ancient Greek name of the Walnut. *Hickory* is an Indian name for *some* of the species of this genus; one of them was known to the Indians by the name of *Pecan* or *Pakan*. Rafinesque applied the barbarous name *Hickoria* to this genus, without describing or limiting it; in so doing he has no higher claims for the adoption of the name than our woodsmen and the aborigines.

Carya Microcarpa.

Small-fruited Hickory. Noyer à petit fruit.

§ I. *Nuts more or less quadrangular.* HICKORY, properly so called.

SMALL-FRUITED HICKORY.

CARYA MICROCARPA. *Foliolis quinis ad septenis, oblongo-lanceolatis serratis promisse acuminatis glabris subtus glandulosis; amentis glabris, nuce subglobosa subquadrangulata, testa tenui.*
Carya microcarpa.—NUTT., Gen. Am., vol. ii. p. 221. DARLINGTON, Flora Cestrica. [Ed. alt.] p. 545.
Juglans compressa. a. *microcarpa.*—MUHL., Catal., p. 88. BART., Flor. Philad., vol. ii. p. 179.
Juglans alba odorata. Balsam Hickory.—MARSHALL, p. 68.

THIS species, allied to *C. tomentosa*, or the Common Hickory, becomes a fine, lofty, spreading tree sixty to eighty feet high, having a diameter of eighteen inches to two feet or more, with an even bark. I first observed it on the banks of the Schuylkill, in the vicinity of Philadelphia; and my friend Dr. Darlington remarks that it is frequent in moist woodlands in the vicinity of West Chester. The nut is of the same form nearly as that of *C. tomentosa*, of a pleasant taste, with a thin shell, but usually small, not much exceeding the size of a nutmeg. It grows, I believe, also in Massachusetts, where I have seen these peculiar nuts. The wood is white and tough, and possessed of most of the good qualities which recommend the ordinary Hickory. This species is remarkable for the smoothness of its leaflets, which, in that respect, approach *C. glabra* or the Pig Nut, but they are everyway larger and less deeply serrate; two or three pairs with a terminal odd one, four to eight or nine inches long and two to three and a half inches wide, oblong-lanceolate, with shallow serrulations, smooth on both sides when fully expanded, except a slight tuft in the axils of the nerves beneath; the under surface sprinkled with minute resinous particles; the lateral leaflets subsessile and rather obtuse at base, the terminal one with a short petiole and attenuated below. Aments three together, upon a common peduncle, slender, nearly quite smooth, scales trifid, the lateral segments ovate, the middle one long and linear; anthers hairy, mostly four, sometimes three or five. Female flowers two or three together, sessile, on a common peduncle; segments of the perianth very long and somewhat foliaceous. Stigma discoid, four-lobed; fruit globose-ovoid, about three-quarters of an inch in diameter; the pericarp thin, with the sutures rather prominent. Nut somewhat quadrangular, with the shell thin.

By the leaves it appears to be allied to *C. glabra*; but the nut, on a small scale, is that of *C. tomentosa*, or the Common Hickory.

PLATE XIII.

A small branch, reduced about one-third. a. The nut.

COMMON HICKORY, (*Carya tomentosa, β. maxima.*) This is a remarkable variety for the great size of its fruit, which are as large as a moderate apple. It grows a few miles from Philadelphia. Mr. Elliott also observed it on the sea-islands of South Carolina.

Carya glabra, (*Juglans glabra*, DU ROI, Harbk., vol. i. p. 335. *J. porcina*, MICH., North. Am. Sylva, vol. i. pl. 38.) Of this there are two varieties, one with globose and the other with turbinate fruit: intermediate forms are also met with, proving them to be no more than varieties.

II. *Nuts even, without prominent angles. Leaflets often numerous.*—PECANS.

Carya angustifolia. Juglans angustifolia, AIT. Kew., vol. iii. p. 361. *J. Pecan*, MUHL. in Nov. Act. Soc. Nat. Scrut. Berolin., vol. iii. p. 392. *J. oliræformis*, WILLD., Sp. Pl. 4, p. 457. A fine, stately tree, formerly *cut* down for the sake of obtaining a single crop of nuts; remarkable for its numerous leaflets and their almost falcate form. In Massachusetts, where it has been submitted to cultivation, it never grows beyond the size of a shrub, being every year more or less cut down by the effects of the severe frosts.

Carya Pecan. Juglans Pecan, WALTER. *J. myristicæformis?* MICH., Sylva, vol. i. pl. 39. This obscure plant of Walter may, perhaps, be nothing more than *C. glabra*. Michaux's plant was unknown to Elliott.

Carya amara. Juglans amara, MICH., Sylva, vol. i. pl. 33.

Observations. According to an experiment published in the "Massachusetts Agricultural Journal," the sap of the Butternut Tree (*Juglans cinerea*) is capable of producing as much sugar as that of the Maple. Four of the trees yielded in one day nine quarts of sap, which produced one and a quarter pounds of sugar.

THE BLACK WALNUT (*Juglans nigra*) is met with as far north as Massachusetts, particularly in the western part of the State, as around Northampton. Mr. Emerson says, "*Juglans nigra* I have found repeatedly as far north as Boston. It is in Middlesex, Worcester, and Norfolk counties, Massachusetts."

On the banks of the Scioto, in Ohio, I have seen a tree of six feet in diameter.

CANDLEBERRY MYRTLE.

Natural Order, MYRICEÆ, (Richard.) *Linnæan Classification*, DIŒCIA, TETRANDRIA *to* OCTANDRIA.

MYRICA. LINN., (in part.)

Flowers unisexual; those of the two sexes upon the same or more commonly upon different plants. *Male* flowers in cylindrical sessile catkins; each flower with four to eight stamens, with the filaments elongated and more or less united at the base; the stamens exserted beyond the borders of the dilated short scale, many stamens in branching clusters nearly without scales at the summit of the catkin; bracteoles *none* in either sex. *Female* flowers in loose, sometimes filiform catkins, with many of the lower scales abortive; scales 1-flowered, the germ naked. Styles two, very long, linear, and acuminate; ovary villous. Drupe 1-seeded, spherical, coated with a grumose waxy pulp. Nut very hard; seed erect; embryo without albumen, the radicle superior. Cotyledons thick and oily.

A genus wholly distinct from *Myrica Gale*, which is common to Northern Europe and North America. The character of lunate scales given to Myrica by Linnæus applies only to the GALE, which therefore constitutes a genus by that name. The rest of our species belong to Myrica. In the Gale, the fruit is a small, ovate, dry nut, with an indurated bracte on either side of it, giving it the appearance of being 3-lobed.

The species of this genus are few, natives of the warmer and colder zones of both hemispheres, growing generally near the seacoast, and are chiefly shrubs, with alternate, persistent, or annual simple leaves, usually more or less serrated or pinnatifid, and be-

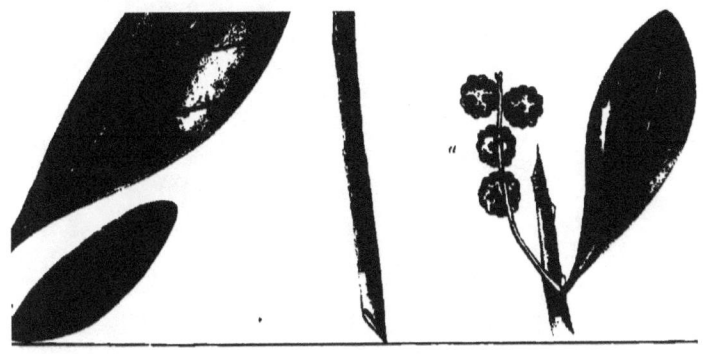

Myrica inodora.

Inodorous Candle-tree　　　　　Cirier inodore

sprinkled with aromatic resinous scales, as are also the scales of the buds. Catkins axillary, expanding early in the year. There are several species in Nepaul in India, and at the Cape of Good Hope. The *M. Faya*, with a 4-celled drupe, and about ten stamens in loose catkins, will, no doubt, constitute a different genus, which I propose to call FAYA AZORICA.

INODOROUS CANDLE TREE.

MYRICA INODORA. *Arborea, foliis lanceolato-ellipticis integris obtusis margine revolutis basi cuneatis subtus vix squamosis plerisque nudis, baccis majusculis albidis.*
MYRICA *inodora*, BARTRAM's Travels in Florida, &c., p. 405.
MYRICA *obovata*, CHAPMAN, MSS.

IN the month of March, 1830, on the borders of the Escambia, in West Florida, I had the fortune to rediscover this fine species of Myrica, so long since described by the amiable and excellent Wm. Bartram, near Taensa Bluff, on the Mobile River, where he remarks, August 5, 1776, "In my excursions about this place, I observed many curious vegetable productions, particularly a species of Myrica, (*Myrica inodora*.) This very beautiful evergreen shrub, which the French inhabitants call the Wax Tree, grows in wet, sandy ground, about the edges of swamps; it rises erect nine or ten feet, dividing itself into a multitude of nearly erect branches, which are garnished with many shining deep-green entire leaves of a lanceolate figure. The branches produce abundance of large round berries, which are covered with a scale or coat of white wax: no part of this plant possesses any degree of fragrance. It is in high estimation with the inhabitants for the production of wax for candles, for which purpose it answers equally well with beeswax, or preferable, as it is harder and more lasting in burning."—Bartram's *Travels*, p. 405–406.

We found it exactly in similar situations as those described by Bartram; and it has also been found in Alabama by Dr. Juet, from whom I have been favored with specimens. It may with propriety be called a tree, though never so large a one as the *Myrica Faya*, or Fayal Myrtle. The stem sometimes attains the thickness of a man's arm, and, like the rest of the genus, it is gregarious and forms stout thickets on the margins of small streams and swamps. The berries are twice as large as those of the common Wax Myrtle. Though the leaves have no perceptible scent, they are not always entirely without the usual scaly resinous glands; they have no serratures, and are about three to three and a half inches long by one to one and a half wide. The bark is of a gray color, inclining to brown. The male catkins are unusually large, as well as the berries, and the leaves, when old, are as stiff as in the laurel. The stamens beneath each scale of the ament are eight, with distinct filaments and monadelphous at base; the summit of the catkin is nearly without scales, and terminates in monadelphous branchlets of stamens, each bearing three or four anthers. The female catkin is loose, and the lower scales empty; the germ is pilose. The wood appears compact, fine-grained, and nearly white. The candles formed of the myrtle wax burn long, yield a grateful smell, and are destitute of the disagreeable scent produced on extinguishing tallow candles. In Carolina, a kind of sealing wax has been made of it, and the root has been accounted a specific in toothache. In Prussia it has been cultivated for the wax.

The Fayal Myrtle (*M. Faya*) is in Fayal the principal article of fuel; it there attains the ordinary height of a peach-tree, with a more erect stem: it produces a considerable quantity of compact, reddish wood. It is also cultivated in rows between and around the orange trees for the purpose of shelter from the cutting sea-breezes, which would otherwise abridge their height and retard their growth. All the gardens of the island require the same shelter for which the evergreen Faya is so well calculated, being perfectly hardy and indigenous to the Azores.

PLATE XIV.

A branch of the natural size.   a. The berry.

PLANE TREE.

Natural Order, PLATANEÆ. *Linnæan Classification*, MONŒCIA, POLYANDRIA.

PLATANUS.* (TOURNEFORT.)

Flowers of one sex, those of the two kinds situated upon the same plant, and each of them disposed in spherical aments on pendulous stalks, producing from two to five upon each. MALE flowers formed of minute thickish bractes; the filaments very short, situated between the bractes. *Anthers* 2-celled, attached to a connectivum broader than the filament, with a peltate summit. FEMALE. *Pistils*, numerous, in pairs. *Ovary* of 1-cell, including one to two pendulous ovules. *Stigmas* 2, long and filiform. *Fruit*, a carpel seated in a tuft of articulated hairs, including one pendulous oblong seed, destitute of albumen.

LOFTY, deciduous-leaved trees, with widely-spreading branches and a dense, broad foliage having a pentangular outline. Natives of Europe, Asia, Northern Africa, and the temperate parts of North America. The species about three. Leaves alternate, palmate; the bud concealed beneath a conical envelop and immersed in the base of the petiole. The young shoots and leaves covered with a deciduous down. The old bark at length scaling off in extensive patches, leaving the trunk smooth.

* The name is from the Greek word *platys*, broad, in allusion to its wide-spread leaves and branches.

Platanus racemosa.

CALIFORNIA BUTTONWOOD, OR PLANE.

PLATANUS RACEMOSUS. *Foliis quinquelobo-palmatis basi truncatis subsinuatis subtus lanuginosis pallidis, lanciniis lanceolatis acuminatis integris, stipulis angulatis fructibus racemosis.*—NUTTALL, Mss. in Audubon's Birds of America, tab. 362.

THIS remarkably distinct species of Platanus is a native of Upper California, in the vicinity of Santa Barbara, where it puts on very much the appearance of our common Buttonwood, (*Platanus Occidentalis.*) As far as I yet know, it is the only species on the western coast of America. It grows probably farther north, but I did not meet with it in the territory of Oregon. It does not appear in this unfriendly climate to arrive at the gigantic magnitude of its Eastern prototype, though it equally affects rich bottom-lands and the borders of streams; but the scarcity of rain, in this climate, which had not for three years been sufficient to encourage the raising of crops, and the consequent disappearance of water in most of the brooks, prevented, no doubt, this subaquatic tree from assuming its proper character in a more favorable soil. At first view it would be taken for the ordinary species, spreading out the same serpentine picturesque limbs, occasionally denuded of their old coat of bark, and producing the same wide and gigantic trunk; but a glance at the leaves, no less than the fruit, would remind the Eastern traveller that he sojourned in a new region of vegetation, and objects apparently the most familiar he met around him, associate them as he would, were still wholly strangers.

The leaves not fully expanded were about four inches wide and the same in length, divided more than half-way down into five sharp-pointed, lanceolate portions, of which the two lower are the smallest: all the divisions are quite entire; two of them in small leaves are suppressed, thus producing a leaf of only three parts. Above, as usual, the surface is at first clad with a yellowish copious down formed of ramified hairs, which quickly falls off and spreads itself in the atmosphere. The under surface of the leaves are, however, always copiously clad with a coat of whitish wool, which remains. The young leaves, clad in their brown pilose clothing, have a very uncommon appearance, and feel exactly like a piece of stout, thick woollen cloth. The branchlets, petioles, and peduncles are equally villous. The *male* catkins are small, less in size than peas, full of long-haired scales, and with unusually small anthers. The *female* catkins are in racemes of three to five in number, with remarkably long styles, being between two and three-tenths of an inch in length, and persistent on the ripe balls. The raceme with the full-grown balls measures nine inches. The tree has, therefore, a very unusual appearance, filled with these very long pendulous racemes, each bearing from three to four or even five balls, at the distance of about an inch from each other. The stigmas are at first of a deep and bright brown.

The wood of this species, as far as I could learn from the American residents at Santa Barbara, is far preferable to that of the common Buttonwood, being much harder, more durable, less liable to warp, and capable of receiving a good polish: it is of a pale yellowish color, like the young wood of the Oriental Plane, and bears some resemblance to beech wood in its texture. In the radiation of its medullary vessels, it resembles the wood of the common species.

PLATE XV.

A branch of the natural size. *a. The carpel.*

The ORIENTAL PLANE (*Platanus Orientalis*) deserves to be planted in the United States as an ornamental tree. It grows to the height of from seventy to ninety feet, with widely-spreading branches and a massive trunk, forming altogether a majestic object. The leaves are more deeply divided and indented than in our common species. A native of the East, where shady trees are not so abundant as in North America, it was celebrated in the earliest records of Grecian history. Xerxes, it seems, (according to Herodotus,) was so fascinated with a beautiful Plane Tree which he found growing in Lycia, that he encircled it with a ring of gold, and confided the charge of it to one of the Ten Thousand. He passed an entire day under its shade, encamping with his whole army in its vicinity; and the delay so occasioned was believed to be one of the causes of his defeat. Pausanius (A.D. 170) mentions a Plane Tree of extraordinary size and beauty in Arcadia, which was said to have been planted by Menelaus, the husband of Helen, and to have been, at the time he saw it, 1300 years old.

Plane Trees were planted near all the public schools in Athens. The groves of Epicurus, in which Aristotle taught his peripatetic disciples, the shady walks planted near the Gymnasia and other public buildings of Athens, and the groves of Academus, in which Plato delivered his celebrated discourses, were all formed of this tree.

The remarkable Plane Tree at Buyukdere, or the Great Valley, mentioned by Olivier, the naturalist, and after him by Poucqueville, Hobhouse, and various other writers, has a trunk that presents the appearance of seven or eight trees having a common origin, which Olivier supposes to be the stool of a decayed tree, and which were all connected at their base. Dr. Walsh, who measured the tree in 1831, found the trunk one hundred and forty-one feet in circumference at the base, and its branches covered a space of one hundred and thirty feet in diameter. The trunk divides into fourteen branches, some of which issue from below the present surface of the soil, and some do not divide till they rise seven or eight feet above it;

one of the largest is hollowed out by fire, and affords a cabin to shelter a husbandman. The tree, if it can be considered a single plant, is certainly the largest in the world. But what renders it an object of more than usual interest is, that M. Decandolle conjectures it must be more than 2000 years old.

The wood of the Oriental Plane, in the Levant and in Asia, is used in carpentry, joinery, and cabinet-making. It is said to make beautiful furniture, on account of the smoothness of its grain and its susceptibility of receiving a high polish.

Concerning our common Plane Tree or Buttonwood, (*Platanus Occidentalis*,) Dr. Darlington remarks, in his "Flora Cestrica," page 542, "It makes a noble shade in front of houses where it has room to develop itself:" and he further remarks, that "the *wood* is not much esteemed, but is occasionally sawed into joists and other lumber." It is beginning to be considerably planted as a shade-tree on the side-walks of the streets in several of the large towns of the United States, and, being seldom attacked by insects, and rarely elevating the pavements, it is exceedingly well calculated for this useful purpose in a climate subject to such ardent summer heats; but, if the Oriental Plane would answer the same purpose, and it is easily propagated, we should not only possess an ornamental but also a *useful tree*, as it regards the wood. The finest specimens of trees of this species, in the vicinity of Philadelphia, are those round the Pennsylvania Hospital, which were planted about the year 1760.

POPLARS.

Natural Order, AMENTACEÆ, (JUSS.) *Suborder,* SALICINEÆ.
Linnæan Class and Order, DIŒCIA, OCTANDRIA.

POPULUS.* (LINN.)

DIŒCIOUS. *Aments* cylindric, with the scales deeply cleft. *Perianth* cup-shaped, oblique, and entire. *Stamens* about eight, (or from thirty to one hundred or more,) inserted on the scale or perianth. FERTILE FLORETS with the scales and perianth as in the male. *Stigmas* three or four. *Capsule* 1-celled, 2 to 4-valved. *Seeds* numerous, comose, with long, soft hairs like wool.

Trees of the temperate and colder parts of Europe and North America, with one species in Asia. The leaves are alternate, roundish, or deltoidly cordate, the petiole, for the most part, vertically compressed toward the summit, and often glandular at the base; the flowers (as in the Willows, to which they are intimately allied) appearing before the leaves.

The Poplars are divisible into two sections or subgenera.

§ 1. Those properly so called, with about eight stamens, and, usually, filiform stigmas.

§ 2. Those with from thirty to one hundred or more stamens, and with broad, dilated, reniform stigmas. POTAMERIA. These are mostly large trees, which affect the banks of rivers, and includes *Populus lævigata,* (*P. Canadensis,* MICH. fil.,) *P. angulata, P. monilifera, P. heterophylla,* (*P. argentea,* MICH. fil.,) and probably *P. candicans, P. balsamifera,* and our *P. angustifolia.*

---

* An old Latin name of uncertain derivation.

NARROW-LEAVED BALSAM POPLAR.

POPULUS ANGUSTIFOLIA. *Foliis ovato-lanceolatis lanccolatisve acutis, superne attenuatis penninerviis concoloribus glabris adpresso-serratis; ramulis teretibus glabris, gemmis resinosis.*

P. angustifolia, TORREY, Lyceum Nat. Hist. N. York, vol. ii. p. 249.
Narrow-leaved Cottonwood, of LEWIS and CLARKE.

As we ascended the banks of the river Platte, in our extended journey to the West, about Laramie's Fork, a northern branch of that extensive stream, we observed scarcely any other tree along the alluvial plains but the present and the Cottonwood; and those were chiefly confined to the islands, a circumstance accounted for by the annual burning of the prairies, which wholly strips the streams of their margin of forest, so that we behold, far and wide, nothing but a vast plain, a sea of grass undulating before the breeze; and the illusion appears more sensible by the fact that the only variation to the scene is produced by the scattered islands of the lofty Poplar, which gives life and variety to the wild and boundless landscape.

The height of this species, which so nearly resembles the Balsam Poplar, may be about sixty to one hundred feet, having a trunk of proportionate diameter, clad, like the Cottonwood, with a rough, grayish bark. Although a brittle and poor wood for almost every purpose, it will, like the Cottonwood of the Mississippi, (*Populus angulata,*) become, of necessity, important for fence and fuel, whenever this country shall become settled, as scarcely any other timber exists in sufficient quantity for economical purposes. When dry, it burns well, but is quickly reduced to ashes.

Whatever may be the immediate uses of the Narrow-leaved Poplar, we must say that, in a country so exposed and arid as the Rocky Mountain region, we felt grateful for the shade and

Populus Angustifolia.

Peuplier baumier à feuilles étroites   Narrow-leaved Balsam Poplar

shelter it so often exclusively afforded us, with the exception of a few insignificant Willows, that frequently associate with it. In short, we rarely lost sight of this tree, which accompanied us to the northern sources of the Platte, presented itself, as usual, on the alluvial banks of the Colorado of the West, along Lewis River, of the Shoshonee, the banks of the Oregon from the Walla-Walla to its estuary, attaining along the banks of this noble stream, and its southern tributary the Wahlamet, an augmention so great as to vie in magnitude with the tall Cottonwood of the Mississippi, and to pass amidst the mighty forests of the West as one of the largest deciduous-leaved trees of the country. We find this species of Poplar also on the banks of the Missouri, on the upper part of the river, from whence it continues uninterruptedly to the valleys of the Rocky Mountains. In the severity of winter, the boughs are collected by the aborigines to support their horses; the beaver likewise feeds upon them by choice.

Early in the spring the balsamic odor of its resinous buds may be perceived far and wide, and the shade of its tall, round, and spreading summit invited hosts of birds to its branches, particularly the large and fine Band-tailed Pigeon, which feeds with avidity on its seed-buds in the spring, affording them an abundant article of food.

The lesser branches are rather tough than brittle, covered with a smooth, yellowish bark. The leaves vary on the same branch, in this respect, that the earliest brought out are the widest; these are generally ovate and acute, (not acuminated as in the Balsam Poplar,) rounded below; afterward come out a set of lanceolate leaves, much longer than the others, often three inches by only one inch wide: these are acute at both ends, and would pass very well for those of a Willow; they are all margined with fine, bluntish, appressed serratures, and beneath they are nearly the same color as above. The footstalks of the leaves or petioles are about three-quarters to an inch in length. The nerves are all faint beneath and pennate, with no appearance of being 3-nerved at the base; the number of these nerves or lateral vessels are twelve to fifteen on a side, at least double the number they are in the Balsam Poplar. The rachis of the female ament is smooth, the germs much corrugated in drying, and the scaly cup of the germ very shallow, not more than half the size of that of the species to which it is so closely allied. Upon the whole comparison, we are satisfied that this is as distinct a species as any in the genus. In the FLORA ROSSICA, vol. i. p. 67, t. 41. B., a leaf is given as a variety of the Balsam Poplar from Altai, very similar to our present species.

PLATE XVI.

A branch of the natural size. *a. The capsule.*

COTTONWOOD, POPULUS LÆVIGATA, *Willd.* (*P. Canadensis*, MICH., Sylva, vol. ii. pl. 95.)

This species, more hardy than the Mississippi Cottonwood, is found on the banks of most of the Western rivers for a considerable distance toward the Rocky Mountains. It is abundant on the borders of the Arkansas, but is nowhere met with in Oregon Territory.

BALSAM POPLAR, (*Populus balsamifera.*) In *Flora Rossica*, vol. i. pp. 67 and 41, is given a figure of a Poplar from Daouria, called the Balsam Poplar, which does not appear to be our plant. The leaves are wider, with shorter petioles. This species stretches into Eastern Siberia, along the river-banks, to the peninsula of Kamtschatka.

AMERICAN ASPEN, (*Populus tremuloides*, MICH., Fl. Bor. Am., vol. ii.p. 243. *P. trepida*, WILLD.) This elegant species, of which the leaves, like the European Aspen, tremble and vibrate in the faintest breeze, is met with west of the Mississippi, in scattering groups, chiefly in the ravines of the mountains, where springs issue out, and in narrow valleys, where considerable groups of this tree may be seen, into the centre of the Rocky Mountain chain; but they here attain a very inconsiderable magnitude, being seldom thicker than a man's arm.

WHITE POPLAR, (*P. alba,*) now commonly cultivated with us for ornament, soon attains the magnitude of an Oak. This wood, remarkable for its whiteness, is generally used in the South of France for wainscoting and flooring houses, and in England it is employed by turners for a host of small articles and utensils of housekeeping.

The wood of various species of Poplar has been employed for dyeing: that of the Lombardy or Italian Poplar boiled in water with wool communicates a very permanent golden-yellow color; other species, as the Black, Virginia Poplar, and Cottonwood, give, according to the length of time employed and the quantity of the wood boiled, various shades of color, as that of nankeen, muse, the color of the Vicunga, with other shades, and answer conveniently for the ground of other colors.

WILLOWS.

Natural Order, AMENTACEÆ, (Jussieu.) *Suborder*, SALICINEÆ. *Linnæan Classification*, DIŒCIA, DIANDRIA, &c.

SALIX.* (TOURNEFORT, LINN.)

The flowers are DIŒCIOUS† or very rarely MONŒCIOUS, disposed in ovoid or cylindric catkins, composed of undivided scales, which are imbricated over each other, and each with the pistils or stamens form a flower; at the base of these scales exists a small glandular body, which is either simple or bifid, and surrounds the interior organs of reproduction. In the *male* flowers there exist from four to five, or even seven to nine stamens; (ordinarily there are only two.) In the *female* flowers the ovary is single, terminated by a bifid style, having usually four stigmas. The capsule or follicle consists of one cell with two valves. The seeds are very numerous and minute, each terminated by a long tuft of hairs or pappus; the radicle is inferior, or in an inverse position to that of the Poplars, to which they are so intimately allied.

THE Willows, numerous in species, are all (with two exceptions in the Straits of Magellan and Peru) natives of the northern hemisphere, and all of them shrubs or trees, some not more than an inch in height above the ground, confined to the highest summits of lofty mountains, others attaining an elevation of

---

* Said to be derived from the Celtic *sal*, near, and *lis*, water.

† A term used by Linnæus to designate a class of plants which have flowers of different sexes on two different individuals, *monœcia*, with two sorts of flowers on different parts of the same plant.

fifty or sixty feet. The wood is usually light and close-grained: the twigs of several species, used by basket-makers, are remarkable for their pliability. The bark of most of the species contains a peculiar vegetable principle, called *salicine*, which, for intermittent fevers, is nearly as serviceable as Peruvian bark; and the down of the seeds has been manufactured into a coarse paper. Most of the species affect wet or humid situations, being common on the immediate borders of brooks and rivers, which they fringe with a luxuriant and agreeable vegetation. The flowers generally precede the appearance of the leaves, and, though not beautiful, they are seen with delight, as the earliest harbingers of our northern spring. The species, more than two hundred in number, present nearly the same general form of foliage, with the margin entire or serrated; the catkins are lateral or terminal, and the stamens are too variable in number to admit of a classification by them; there are some with only one, a good many with two, others with three, four, five, or even nine or ten. The capsules, though small, afford the best traits of specific distinction. The leaves put on various appearances, and even outlines, in the progress of their growth, and the Willows justly rank among botanists as the most protean and difficult family of plants to discriminate in the northern hemisphere. In our tour across the continent we have met with some remarkable species, four or five of which become trees: we shall offer descriptions of the whole, but only give figures of those which rank among arborescent species.

LONG-LEAVED WILLOW.

SALIX SPECIOSA. *Foliis longissimis lanceolatis serrulatis promisse acuminatis glabris subtus glaucis junioribus pilosus; amentis serotinis 6–9-andris, germinibus pedicellatis lanceolatis acuminatis glabris, squamis lanceolato-oblongis sericeis, stigmatis lobis bifidis.*

SALIX LUCIDA, HOOK. (as it regards the Oregon plant.)—Flor. Bor. Am., vol. ii. p. 148.

No Willow on the American continent presents so remarkable and splendid an appearance as the present; the effect of which is produced no less by its magnitude than the size and beauty of its foliage. Its aspect is that of a large Peach Tree, with the leaves and their stalks from five to eight inches long by an inch to an inch and a half wide; beneath, when adult, they are glaucous, like those of the River Maple. The summit is tufted and spreading, and the tree attains the elevation of about twenty to thirty feet, with a trunk of twelve to eighteen inches in diameter. When in full bloom, which is with the first expansion of the leaves, in May, the numerous and large bright-yellow catkins, loading the branches, emulate the finest Acacia of New Holland; they are also agreeably fragrant, and attract swarms of wild bees and other insects, in continual motion among their waving branches. We have seen this noble species nowhere in such perfection as along the banks of the deep Wahlamet and the wide Oregon, whose numerous islands are almost exclusively decked with this imposing Willow, which continues to the Blue Mountains, and along the neighboring streams as far east as the river Boisée. As we sailed along the smooth bosom of these extensive streams, for many miles we never lost sight of the Long-leaved Willow, which seemed to dispute the domain of the sweeping flood, fringing the banks of the streams and concealing the marshes entirely from view; at every instant,

Salix speciosa.

when touched by the breeze, displaying the contrasted surface of their leaves, above of a deep and lucid green, beneath the bluish-white of silver: the whole scene, reflected by the water and in constant motion, presented a silent picture of exquisite beauty. Immediately behind this foreground of spreading Willows arose, in the first rank of the legitimate forest, the lofty Poplars we have already described, succeeded by the majestic Oaks and Maples, while the distant hills to their summits were impenetrably hid by the vast towering Pines and Firs, which, mingling as it were with the clouds, close in the rest of the landscape with funereal grandeur.

This species is related to the Lucid and Bay Willows, and the buds have something of the same aromatic exudation; the serrulations and the base of the leaf are also equally glandular. The bark of the trunk is rough and divided, the twigs smooth and shining, of a yellowish brown. The leaves, at first green on both surfaces, are, before expansion, clothed with long, brown, loose, parallel hairs, which disappear with the progress of their growth; at length they become silvery and glaucous beneath: they are finely serrulated, acute at both extremities, with the points very much attenuated. The stipules are semicircular and broad, serrulated on the margin. The flowers come out with the expanding leaves from lateral buds, containing, in the male, also two or three leaves, so that the catkin appears as a pedunculated spike. The stamens are very long, from five to nine in a scale, with filaments which are hairy toward the base, and, as well as the hairy, broad, cuneate, serrated scales, are of a bright, golden yellow.

The female plant, at the time of flowering, appears to have smaller leaves than the male, and those on the branches which bear the catkin are green on both surfaces for a considerable time; they are also but little acuminated; five or six leaves, with their appropriate stipules, grow out on the same branch, which terminates in a female spike. The scales of the catkin or spike are oblong or lance-oblong, and less hairy than in the staminiferous catkin. The germ is lanceolate, pedicellate, and smooth, acuminated, and terminated by a short, bifid style, with two pubescent, bifid stigmas. The capsule is likewise smooth, and contains seeds with a very long pappus, as abundant almost as on a seed of cotton.

The wood is whitish and close-grained, and might probably be employed for the same purposes as that of the White Poplar, but the nearly uninhabited state of the country in Oregon prevents the possibility of making any useful experiments. As an ornamental and hardy tree, however, it stands pre-eminent among all its fraternity, and well deserves to be introduced into pleasure-grounds, where it would be perfectly hardy as far north as New York, or in any part of Great Britain.

PLATE XVII.

A twig and leaf of the natural size. a. *The female catkin.* b. *The male catkin.* c. *The male flower and scale of the catkin.* d. *The open capsule.*

Salix pentandra.

Long-leaved Bay Willow.   Saule laurier.

LONG-LEAVED BAY WILLOW.

SALIX PENTANDRA. *Foliis ellipticis acuminatis serratis glabris, petiolis superne glandulosis, amentis serotinis pentandris, germinibus lanceolatis glabris.*—WILLD., Sp. pl. 1. c. VAHL., in Flora Danica, tab. 943. HOST., Sal. Austr. 1, t. 1, f. 2. Eng. Bot., t. 1805.
SALIX PENTANDRA. *Foliis serratis glabris, floribus pentandris.*—LINN., Hort. Cliffort., p. 454, et Sp. pl., p. 144. Flora Lapponica, p. 370, t. 8, fig. 3. GMELIN, Flora Sibirica, vol. i., p. 153, t. 34, fig. 1.
Salix foliis glabris, ovato-lanceolatis; petiolis glandulosis; floribus hexastemonibus.—HALLER, Flora Helvetica, No. 1639.
Salix montana major; foliis Laurinis.—TOURNEFORT, Institutes Rei. Herb., p. 591.
Salix foliis laureo sive lato glabro odorato.—RAII., Hist., p. 1420.
SALIX PENTANDRA, β CAUDATA. *Foliis longissime acuminatis attenuatis; ramulis junioribus hirsutis; capsulis rugulosis opaci.*

WE met with this species, hitherto wholly European, in the very centre of the North American Continent, by streams in the valleys of the Rocky Mountains, toward their western slope, in Oregon, and also the Blue Mountains of the same territory. It is true, the present variety is a remarkable one. The uppermost leaves on the adult branches are about five or six inches long, one-half of which distance, or about two and a half to three inches, may be considered as the slope of the summit and the point of the leaf; the buds and last-developed branches are also hirsute. This variety, like its prototype, becomes a tree fifteen to twenty feet high, of a bright, but not deep, and rich green. The flowers are fragrant, produced after the appearance of the leaves, and the capsules in our plant are slightly corrugated, even when ripe and open, and do not shine as in the common Bay Willow. The leaves have the same odorous glands, and the bark of the branches is smooth, shining, and of a brownish-yellow color. This tree is met with in mountainous situations by streams, in all the northern parts of Europe,—in Britain, France, Switzerland, Sweden, Lapland, and throughout Siberia and Russia. Its branches are too fragile to be employed for any economical purpose, and the wood decrepitates in the fire. The leaves, which are fragrant from the resinous glands of their margin, however, furnish a yellow dye, and the abundant down of its seeds, in some of the northern countries, is used with success as a substitute for cotton, mixed with a third part of the true material.

According to Loudon, it is one of the most desirable species of the genus for planting in pleasure-grounds, on account of the fine display made by the blossoms, their abundant fragrance, the shining, rich, deep green of the leaves, and the comparatively slow growth and compact habit of the tree. It is also one of the latest flowering Willows, the blossoms seldom expanding till the beginning of June.

PLATE XVIII.

A branch of the natural size. a. The capsule.

WESTERN YELLOW WILLOW.

SALIX LUTEA. *Foliis ovato-lanceolatis acutis leviter serrulatis glabris, stipulis lunatis, amentis præcocibus brevibus cylindraceis, germinibus pedicellatis glabris, ramis luteis.*

THIS is a smallish arborescent species of Willow, which inhabits the Rocky Mountain region, and proceeds westward to the banks of the Oregon, but is nowhere common. It attains to the magnitude of a small tree, and at first glance appears

Salix lutea.

Western Yellow Willow.   Saule jaune

nearly allied to the common yellow-twigged Willow (*S. vitellina*,) but it never grows so large, and is, after the very first evolution of the leaf-buds, perfectly smooth, with none of those tufts of hairs which in that species appear at the interior base of the bud; it likewise possesses stipules, and bears short, close catkins, with smooth capsules, and the serrulations of the leaves are not glandular, indeed scarcely visible, the leaf appearing to have a thin and often almost entire diaphanous margin.

It is at the same time a very elegant and distinct species, remarkable for its smooth bright-yellow branches, and pale green, rather small, lucid leaves. It is well deserving of cultivation for the sake of variety, and is undoubtedly hardy from the climate it inhabits. The leaves are rather narrow, more lanceolate than ovate, acute, or somewhat acuminate, about an inch and a quarter long, and less than half an inch wide. The minute serratures or serrulations are rather crowded, but very minute and shallow, and not in the least cartilaginous. The young leaves before evolution are silky and white beneath, but the pubescence wholly disappears with their expansion.

PLATE XIX.

A branch of the natural size with fertile catkins.   a. The capsule.

HOOKER'S WILLOW.

SALIX HOOKERIANA, (Barratt, Mss.) *Ramis valde robustis pubescentibus, junioribus dense tomentosa-lanatis, foliis late obovatis fere subrotundatis rigidiusculis serratis margine planis supra (adultis) nudiusculis subtus tomentoso-lanatis, stipulis [semicordatis,] amentis cylindraceis crassis, squamis longe densissime lanatis, ovariis longe stipitatis lanceolatis glaberrimis, stylo breviusculo, stigmatis lobis integris.*—HOOK., Flor. Bor. Am., vol. ii. p. 145, t. 180.

This small tree, nearly allied to *S. cinerea*, we found of common occurrence in the lower part of the territory of Oregon toward the sea, particularly frequent on the margins of ponds and in wet places near the outlet of the Wahlamet, where it attained the height of eight or ten feet. It was also found in the territory of Hudson's Bay, near the Grand Rapids of the Saskatchewan, by Douglas. The leaves are remarkably protean in their form, sometimes wholly soft and villous on both sides, more commonly so beneath, the nature of the pubescence also varying till it at times resembles in appearance and to the touch the most brilliant velvet; the old leaves are generally obovate, smooth and shining above, often nearly so beneath, and then of a bluish-green or glaucous hue; they are, moreover, both acute and obtuse, sometimes even rounded at the extremity, and are generally among the largest, or at least widest, of Willow leaves. The stipules on some branches are very conspicuous, circular, and serrated. The male aments are extremely pilose in an early stage, as in our *S. conifera*, to which this species makes some approach; the stamens are two to each scale. The capsule is ovate-lanceolate and very smooth, the style elongated, and the stigmas two and undivided.

The twigs of this species are flexible, but the wood is too

small for almost any economical employment. The old wood of the trunk of *S. caprea* or the Sallow, so much like the present species, is much esteemed by wood-turners.

BLUNT-LEAVED WILLOW.

SALIX FLAVESCENS. *Foliis oboratis sublanccolatis flavescente tomentosis integris demum glabris, stipulis parvulis subsemicordatis denticulatis, amentis præcocibus, capsulis lanceolatis sericeo-pubescentibus, stigmatibus villosis.*

WE met with this shrubby species in the range of the Rocky Mountains, and we are doubtful if it is not also an inhabitant of Europe. It agrees very nearly with the Obtuse-leaved Willow of Willdenow, (*S. obtusifolia,*) of which we have seen no specimens, and which is apparently a very obscure species, said to be a native of the Lapland Alps. Ours is a large shrub, with much of the aspect and general character of the Gray Willow, which is indeed the type of a group of kindred Willows. For a good while the leaves still remain downy, particularly on the under surface, which is tinged also with pale yellow. The leaves, when old, are about one and a half inches long, three-quarters of an inch wide or more, wedge-shaped at the base, obovate or oblanceolate, quite smooth, entire apparently, yet the stipules are denticulate. The branches are brown or dark purple. The catkins are short and cylindric, the scales blackish, hairy, and obtuse, the capsules white and silky; the style is distinct, and the four stigmas long and pubescent.

POND WILLOW.

SALIX STAGNALIS. *Foliis oblongo-lanceolatis obtusis integerrimis basi cuneatis subtus pubescentibus, stipulis nullis, amentis coætaneis, capsulis lanceolatis acuminatis sericeis, squamis sublanceolatis.*

This is another species indigenous to the banks of the Oregon, related to the same section with the last. The twigs are dark brown and slender, and the leaves scattered beneath with a minute brown pubescence, which communicates a somewhat rusty appearance to the leaves; they are about one and a half inches long and one-half to three-quarters of an inch wide; in the bud they are covered with long silky hairs. The capsules are remarkable for their great attenuation and length; the scales of the ament are oblong-lanceolate and hairy. We have not seen the male plant, and our account is therefore imperfect.

VELVET WILLOW.

SALIX CUNEATA. *Foliis lanceolatis oblongisve apice subserrulatis acutis, basi cuneatis glabris subtus holosericeis, stipulis minutis plerisque nullis, amentis coætaneis elongatis, capsulis ovatis sericeis, stigmatibus brevibus glabris.*

This beautiful Willow we found growing in clumps near the rocky margin of the Oregon at its confluence with the Wahlamet, attaining the height of six or eight feet. The branches are slender, and, according to their age, vary much in appearance, at first villous and downy, but at a later period brown, and sometimes quite blue, with a glaucous bloom. The leaves, with their short petioles, measure about two inches, and are about half an inch wide, pubescent above, at length nearly smooth, and deep green, but always clad beneath with a whitish close tomentum, producing all the brilliant display of the finest velvet. The male flowers we have not seen. The female catkins are rather long, loose, and subcylindric, often accompanied at the base by two or three leaves, and come out when the leaves are considerably grown. The capsules are silky, short, ovate, and acute; the style short, and the stigmas four and smooth. The scales of the catkin are brown and oval, somewhat hairy, and much shorter than the capsules. In the narrow-leaved varieties, the leaves appear almost wholly entire. The broader-leaved plants bear some resemblance to the Gray Willow, but the serrulations are minute and the stipules very small, or wholly wanting.

WESTERN POND WILLOW.

SALIX MACROCARPA. *Foliis lanceolatis angustatis subintegerrimis utrinque acutis subacuminatis demum glabris subtus glaucis, stipulis obsoletis, amentis coætaneis diandris, capsulis ventricosis caudatis glabriusculis pedicellatis, stigmatibus subsessilibus quadrifidis.*

THIS species, like our Pond Willow (*S. grisea,*) to which it is closely related, is found forming clumps in wet places where the water is stagnant,—situations which it always seems to prefer to the banks of running streams. It attains the height of three or four feet. The branches are smooth and brownish black, sometimes glaucous or whitish. The leaves (about two to three inches long and half an inch wide) are at first covered with a brownish silky down which disappears with age, when they become dark green and bluish white, or glaucous beneath; they are usually very acute, and mostly entire. The catkins are small and oblong, with two or three leaves at the base of the pedicel; the scales of the stamens are small and blackish, oval and obtuse, somewhat hairy. The female aments have very short pedicels, and produce at the base about two leaves; the scales are narrow and linear; the capsules pedicellated, somewhat villous, but at length nearly smooth, ventricose-lanceolate, with long points, and nearly sessile stigmas.

SOFT-LEAVED WILLOW.

Salix sessilifolia. *Foliis lanceolatis sessilibus acutissimis apice subserrulatis villosis mollibus, stipulis nullis, amentis scrotinis diandris elongatis terminalibus; germinibus lanatis, capsulis lanceolatis demum subglabris, stylo profunde bipartita stigmatibus bifidis.*

This beautiful and very distinct species of Willow formed dense tufts on the rocky borders of the Oregon, at the confluence of the Wahlamet, attaining to the height of about six to eight feet, and when in flower appeared as showy as a *Mimosa*. It is remarkably leafy, and the leaves are hoary, with a rather long and somewhat copious pubescence, which communicates a softness to the touch equal to that of velvet; whether the leaves afterward become more smooth or not I cannot say, but think it probable. Different from almost every other Willow I have seen, the catkins each terminate so many small, leafy branches about two inches long, exclusive of the catkin or spike, which is itself about one and a half inches, in both sexes attenuated at the base; these branchlets, as well as the leaves, are whitish, with soft hairs, particularly the former, but still the green color of the leaf predominates; the points of the leaves are somewhat rigid, sharply acute, and, unlike most other species, they are destitute of petioles or footstalks; the scales of the ament are oval and unusually conspicuous, more hairy in the staminiferous flower. The stamens are two to a scale. The capsule is pubescent and lanceolate, at length nearly smooth; the stigmas four, and rather long. No vestiges of stipules appear in any of our numerous specimens; the older branches are dark brown.

It is difficult to decide on the affinities of this very distinct species, which at a little distance resembles a *Protea* or *Leucodendron*, the leaves being equally gray and silvery, with soft hairs, which are so equally distributed on either surface as to obliterate the presence of the veins and render both surfaces almost similar. It appears, in some respects, to resemble *S. arenaria*, the Sand Willow, but the late appearance of the aments and their remarkable disproportion are almost without a parallel.

PROSTRATE WILLOW.

SALIX BRACHYCARPA. *Foliis ovatis lanceolatisve acutis subsessilibus integerrimis cinereo-pubescentibus subtus incanoxillosis, stipulis nullis, amentis coxtaneis brevibus glomeratis, capsulis ovatis abbreviatis tomentosis, stigmatibus subsessilibus.*

This singular prostrate and dwarf Willow we met with in the Rocky Mountain range, on the borders of Bear River, a clear, rapid brook cutting its way through basaltic dikes to the curious lake of Timpanogos, in New Mexico. This locality is likewise famous for the numerous seltzer springs, so strongly impregnated with carbonic acid as to sparkle and effervesce like champagne. Our hunters called them the "Beer Springs," and, for a day and a half that we spent at this memorable place, the waters afforded us a most delicious treat during the warm weather, in those arid plains. In an open, marshy situation, on the margin of the river-forest, grew an abundance of this curious, depressed, and hoary shrub, which has somewhat the aspect of the European Sand Willow, (*S. arenaria;*) but it is much more dwarf, with the leaves smaller and always hoary with pubescence. The stem branches from the base, only rising four or five inches above the surface of the earth, but with many diffuse, tough, woody branches, which spread out into a circle of a foot or more. The root-stock is woody and thick; the branches full of cicatrices, pubescent, but brown beneath, thickly covered with small leaves, which in some plants are elliptic-ovate, in others oblong-lanceolate, all very entire, nearly sessile, and acute, from half an inch to an inch long, about three lines wide; above always gray with pubescence, but beneath rather whitely villous; some of the lowest small leaves are smooth on the upper surface. There are no stipules in any of my specimens. The male flowers I have not seen. The fertile catkins are short and somewhat clustered, not cylindric, few-flowered, the capsules oblong-lanceolate, and short, villous, with appressed hairs, not densely lanuginous, as in *S. arenaria,* terminated by a short, slender style and four short stigmas.

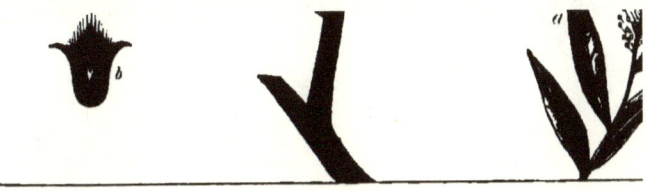

Salix argophylla.

Silver leaved Willow   *Saule à feuilles argentées*

SILVER-LEAVED WILLOW.

SALIX ARGOPHYLLA. *Foliis lineari-sublanceolatis acutis sessilibus integer rimis utrinque argenteo-sericeis, stipulis obsoletis, amentis serotinis diandris, capsulis villosis lanceolatis.*

IN our devious progress to the West, we at length approached one of the branches of the Oregon, the river Boisée, toward its junction with the Shoshonee; its banks were not fringed with a belt of forest, but so stripped of every character of an alluvial nature, that when we suddenly approached it there appeared no break in the plain, and the clear and rapid flood shot through a deep, perpendicular chasm of columnar basaltic rocks. We descended toward its brink and pursued our path along its obstructed banks, climbing over fearful rocks and along the margins of impending precipices: night approached without any cessation of our incessant toil. At length we hailed with satisfaction a small portion of the river-bottom, almost covered by tall bushes of a remarkable, silvery appearance, which proved to be the subject of our present examination, a very curious species of Willow, which, with a kind of Mock Orange, (*Philadelphus,*) composed nearly all the conspicuous vegetation of this sterile chasm. At length our new Alnus, (*A. Oregona,*) the Narrow-leaved Balsam Poplar, and the Long-leaved tree Willow, gave evidence of an ample, alluvial plain and the proximity of the Great Shoshonee.

This species becomes a small tree from twelve to fifteen feet in height, as silvery and white as the *Leucodendron argenteum;* the branches are brown, but the twigs are hoary with villous hairs. The leaves are very much crowded, soft, with whitish, shining, silky down, so abundant on either side as wholly to hide the veins and nearly the midrib; they are also nearly without footstalks, entire on the margin, of a narrow, linear outline and sharply acute, with a distinct, bristly point, one and a half to two inches long, and only about three lines wide. Stipules small and linear, seldom seen. The aments come out late with the leaves, and the flower-branches produce from four to seven leaves. The male ament is small and narrow, with the scales lanceolate and villous; the female aments are oblong, the capsules lanceolate and villous.

The wood of this species, though small, is very white, smooth, and close-grained: the species must also be hardy, well worth cultivating, and would then probably grow to a considerable size. We perceive no affinity that this species bears, except perhaps to the *S. angustifolia* of the borders of the Caspian, from which at the same time it is probably very distinct.

PLATE XX.

A branch of the natural size. a. The male catkin. b. The capsule.

LONG-SPIKED WILLOW.

SALIX MACROSTACHYA. *Foliis lineari-sublanceolatis elongatis remote serrulatis acutis utrinque argento-sericeis, stipulis semiovatis deciduis, amentis longissimis præcocibus, capsulis lanceolatis demum glabriusculis.*

WE met with this species sparingly on the banks of the Oregon. It bears a strong resemblance to the last, without, however, being any way intimately allied to it. It forms a slender bush, remarkable for its white and silvery pubescence, which appears as soft and glossy as velvet: the twigs are also pubescent. The leaves are two to three inches long, two or three lines wide, distinctly serrulated, and nearly sessile, with the veins conspicuous through the pubescence. The female catkins, with their peduncles, are three or four inches in length; the capsules are nearly sessile, and at length but slightly pubescent. In this species there are distinct stipules on the young branches.

RIVER WILLOW.

SALIX FLUVIATILIS. *Foliis linearibus utrinque acuminatis, sublanceolatis spinuloso-serratis demum glabris concoloribus, stipulis nullis, amentis scrotinis pedunculatis villosis diandris, squamis oblongis, fructibus lanceolatis glabris pedicellatis, stigmatibus sessilibus.*

This species lines the immediate border of the Oregon, a little below its confluence with the Wahlamet, attaining the height of about six feet or more. We believe this is also the same Willow that we mistook for the Long-leaved species of Pursh and Muhlenberg, (*Salix longifolia,*) which so commonly lines the banks of the Missouri and Mississippi, and which often forms the exclusive growth of the small islands and sandbars, preparing these waters, recovered from the flood, for a superior growth of trees; and they are also accompanied and succeeded commonly by the Cottonwood (*Populus Canadensis.*) We met with this species likewise on the banks of Lewis River of the Shoshonee, accompanying our Long-leaved Willow, both of which continue almost uninterruptedly to occupy the banks and bars of all the Western streams to the Oregon, and proceed along that river to the borders of the Pacific.

In this remarkably fluviatile species, the leaves of the very young plants are somewhat pinnatifid, and at all times the serratures, rather distant, are sinuated and very sharp or spinulosely acute. The branches are brownish and very full of leaves. The leaves, two to three inches long, are seldom more than two lines wide, except in the young shoots, when they are twice that breadth. The branches producing the male flowers are as short as usual, bearing only three or four small leaves; the catkins are narrow, solitary, and rather short, and come out with the opening of the leaves. In the female plant the inflorescence is similar with that of the male; but there is also produced a later growth of catkins, which terminate proper divided branchlets. The scales of the female catkin are oblong and densely bearded below; the germ is smooth, with four sessile stigmas as in *S. longifolia*. The young leaves are at first somewhat hoary and pubescent, with minute hairs; the young plants have also often pubescent foliage.

We have met with the *Salix longifolia* on the banks of the Arkansas, and it greatly resembles the present species, but differs in producing distinct stipules, in the minute serratures, and above all in the pubescent capsule and elongation of the catkin. The leaves are also generally broader, and it is said to grow only about two feet high.

SLENDER WILLOW.

SALIX EXIGUA. *Foliis linearibus utrinque acutis subintegerrimis sericeis, stipulis nullis, amentis serotinis elongatis, capsulis lanceolatis sessilibus, demum nudiusculis.*

This species is also a native of the Territory of Oregon, and grew with the preceding, which it strongly resembles: it is, however, a smaller species; with still narrower leaves, at all times more or less gray and silky; the serrulations are mostly wanting, though very minute ones are sometimes seen: the capsules are smaller, and not pedicellated. The male plant I have not seen. The branches are reddish brown and smooth.

ROUND-LEAVED WILLOW.

SALIX ROTUNDIFOLIA. *Foliis subrotundis ovalisve minutissime serrulatis vel integriusculis utrinque glabris concoloribus, petiolis ramulisque pilosis, stipulis maximis dilatato-cordatis membranaceis glanduloso-serratis, amentis coætaneis diandris lanuginosis, capsulis glabris lanceolatis, stylis elongatis.*
β OVATA. *Foliis ovatis acutis dense serrulatis.*

THIS dwarf and remarkable species of Willow, attaining about two feet in height, was obtained in the gorge of a lofty alpine ravine, through which we fruitlessly endeavored to pass. It was in the month of July, and the perpetual snows which still covered the mountains mingled their outline with the skies. On the skirt of this gelid region grew our present subject. We named this scene of toil and disappointment Thornburgh's Pass, (or rather ravine, as no passage was practicable,) from the man who undertook to be our guide. It was in the central chain of the Rocky Mountains, and near to the pass of the Shoshonees, which the following day we attained.

We know of no species with which we can compare this Willow. The older branches are brown, smooth, and full of cicatrices left by leaves that have grown near together, giving the plant a stunted appearance; the petioles are about half an inch long, with the younger branches hairy; the younger leaves are also somewhat so on the midrib. The leaves are nearly round, from one and a half to three inches wide, and about the same in length, though some of the later-produced leaves are ovate and sometimes even acute; both sides are equally green, the margin in the ovate leaves elegantly and very closely serrulated, but in the round leaves the serrulations are often nearly obliterated. The stipules are very large, wide, and heart-shaped, finely serrated with glandular points; at length they become membranaceous and deciduous. The male aments are oblong, large, and sessile, the scales blackish and ovate, producing copious white hairs longer than their whole length; the filaments are very long. The female aments grow on thick stalks, and have the scales also very woolly; the capsules are smooth and ovate-acuminate; the style is long, terminated by four stigmas. The *ovate-leared* variety was collected by the late Dr. Gairdner, on the hills of the Wahlamet.

MINUTE WILLOW.

SALIX NIVALIS. *Foliis ovalibus subspathulatis integerrimis glabris obtusis, subtus concoloribus reticulatis, amentis serotinis pedicellatis paucifloris, germinibus ovatis sericeis squamæ glabræ retusæ longioribus, stigmatibus subsessilibus.*
SALIX *nivalis.*—HOOKER, Flor. Bor. Am., vol. ii.

THIS elegant and very diminutive shrub, merely about two inches high, was obtained in the same alpine ravine as the former, but on a plain elevated to the very line of perpetual snow, at the height of about 10,000 feet above the ocean-level. The stems are smooth, brown, and woody, sending out very few short branches terminated by scanty tufts of smooth coriaceous leaves, two or three lines wide and about four lines long. The male plant we have not seen; the female catkin is produced upon a pubescent pedicel nearly its length; the flowers are about six, sessile, and the subtending scales smooth, retuse, and nerved. The germ is short-ovate and silky, terminated by a sessile or almost sessile quadrifid stigma.

This species is allied to the *Salix myrtilloides,* but at the same time perfectly distinct.

PLATE XIX.

A fertile plant of the natural magnitude. *a. A leaf.* *b. The young capsule and its scale.*

Salix Melanopsis.

DUSKY WILLOW.

SALIX MELANOPSIS. *Foliis lineari-lanceolatis serrulatis glabris basi attenuatis vix petiolatis, stipulis nullis, amentis coætaneis diandris, squamis striatis obtusis, capsulis glabris lanceolatis subsessilibus.*

THIS species, which I have called Dusky Willow from the dark appearance it assumes in drying, we met with at our station called Fort Hall, in the plains of the Rocky Mountains, on the alluvial lands of Lewis River of the Shoshonee, not growing in masses, but scattered over the banks of the river in the more elevated situations, and there attaining the magnitude of a small tree twelve to fifteen feet high, with a spreading summit, and when in flower forming a very elegant object. It is closely related to the *Triandrous Willow* of Europe, (*Salix triandra*,) but still sufficiently distinct. It never, like that species, becomes a considerable tree, but more resembles in its magnitude and mode of growth our common Black Willow, (*S. nigra.*) The wood is white and close-grained as in that species: the young branches appear blackish-brown in a dead state, and the young leaves appear also very dark; they are about an inch and a half long, and about three lines wide, attenuated at the base, so as to present no distinct petiole, acute above, with minute serratures; there are no hairs at the base of the bud, as in *S. triandra.* The male flowering branchlets are provided with five or six leaves; the catkins linear and elongated, with dusky pubescent scales, marked each with about five striatures. The female catkins are rather short, and the fruit smooth, with sessile stigmas, as in other species of this particular group.

PLATE XXI.

A branch of the natural size.   *a. The scale.*   *b. The stamen.*   *c. The germ.*

The Salix triandra so nearly allied to the present species, becomes a tree thirty feet high, and is frequently planted in osier grounds for the basket-maker; but the best kind planted for this purpose is the *Osier*, properly so called, (*S. viminalis*,) which might be propagated in almost every part of the United States.

BLACK WILLOW.

Salix nigra, *Mich.*, Arb., vol. ii. plate 125, fig. 1. A variety of this tree occurs in South Carolina and Florida, in which the leaves are villous and the scales of the ament densely lanuginous. In the herbarium of Mr. Schweinitz it was marked, on the authority of Elliott, as a species *S. subvillosa.*

This tree, a native of all the States from New England to Florida, and west nearly to the foot of the Rocky Mountains, is one of the few native species which becomes a tree, attaining the height of fifteen to twenty-five feet, with a diameter of ten to fifteen inches. It affects the banks of rivers and lakes, and generally grows near to the water, in moist, occasionally overflowed, situations. In the warmer parts of the States it puts on a handsome appearance, particularly when in flower, but has the defect of branching almost from the base, and leaning in a posture more wild and picturesque than beautiful. It has a dark-colored rough bark, and the branches are brittle at the base; the male catkins coming out with the leaves are very elegant, and attractive and useful to bees. In the southern parts of the Union, according to Mr. Elliott, the stems when found sufficiently large are employed for the timbers of boats, and are found to be light and durable.

WHITE WILLOW.

The WHITE WILLOW (*Salix alba*, LINN.) appears almost naturalized in many parts of the United States. It is particularly abundant in the western interior of New Jersey along the borders of the Wallkill. It grows rapidly, becoming a stout tree in a few years, and in Europe it is considered the most valuable timber tree of the genus; it produces a white, close-grained wood, capable of receiving a polish; it also affords fence-wood, fuel, and bark for the tanner nearly as good as that of the Oak.

The uses of Willows and Willow-wood in Europe are very numerous. It is generally a close, fine-grained, white wood, capable of taking a smooth and equal polish, and remarkably light withal. The osiers are very extensively used for all kinds of basket-work; and, as Virgil remarks, the shepherd sits beneath its shade, while it affords fence for his field, browsing for his flock, and honey for the bees:—

"Salices, humilesque genistæ,
Aut illæ pecori frondem, aut pastoribus umbram
Sufficiunt, sepemque satis et pabula melli."
GEORGICON, II.

MAGNOLIA.*

Natural Order, MAGNOLIACEÆ, (Jussieu.) *Linnæan Classification,*
POLYANDRIA, POLYGYNIA.

Calyx of three deciduous petaloid sepals. *Corolla* of from six to twelve petals. *Stamens* numerous as well as the pistils. *Carpels* disposed in an imbricated cone, 1 to 2-seeded, opening by the dorsal suture. *Seeds* pulpy, red, suspended, when ripe hanging out of the carpel by a long umbilical thread composed of spiral vessels.

Trees and shrubs with large, entire, alternate, deciduous leaves, and solitary, terminal, large, and usually odoriferous flowers. Chiefly natives of North America, China, and Japan.

LARGE-FLOWERED MAGNOLIA.

MAGNOLIA GRANDIFLORA, *Linn.* (MICH., Sylva, plate 51.)

IN the neighborhood of Savannah, in Georgia, and near New Orleans, this splendid tree often presents an almost equal, smooth, columnar shaft of sixty to eighty feet elevation, and

---

* Named by Linnæus in honor of Pierre *Magnol*, a botanist of Montpellier.

attains the height of one hundred feet or upward, with a graceful, high, and spreading summit. On the trunk of this species, near Savannah, I observed large quantities of the parasitic air-plant, *Epidendrum conopseum,* and it appeared there to grow on no other tree. According to Wm. Bartram, who saw the species so abundant in his tour in Florida, its summit forms a perfect cone, rising from a straight clear trunk, resembling a beautiful column; and, from its dark foliage "silvered over with milk-white flowers," it is seen at a great distance. The succession of flowers is also long continued; in favorable situations from May to August. Though confined very much to the neighborhood of the sea-coast, it extends westward in Georgia as far as Milledgeville; and I met with it in Alabama, on the banks of Utchee Creek, about twelve miles from Columbus in Georgia, and afterward in other parts of that State down to West Florida. It is known to the Creeks by the native name of *Toola.*

In the new edition of Duhamel, we have the following account of its introduction into France. There is at Maillardière, about five miles from Nantes, a fine Magnolia, which was brought from the banks of the Mississippi, in 1732, and planted in a poor soil. It grew there neglected for more than thirty years, till M. Bonami, a physician of Nantes and professor of botany there, recognised this beautiful tree to be the *Magnolia grandiflora;* and at the meeting of the States of Bretagne in September, 1760, in Nantes, he presented to the Princess of Rohan-Chabot a fine branch of this Magnolia in flower, which became a subject of conversation and interest to all assembled. Louis XV. possessed several small plants of this species in his garden at the Petit Trianon, but they did not thrive; and, having heard of a Magnolia thirty-five to forty feet high, which every year was covered with fine flowers of a delicious perfume, he sent two of his gardeners to ascertain if it was possible to transport this tree to Versailles, and, above all, should they do so, if it would be certain to grow. They saw the tree, and, being of opinion that it would not survive removal, it was suffered to remain in its place. It was at that time from thirty-five to forty feet high; but, during the troubles of the civil war of La Vendée, it was mutilated, and lost most of its branches. Afterward, the burning of the house near which it was planted having damaged its head, the branches were cropped down to the trunk, and it again shot out with vigor, but the young shoots, not having had time to ripen, were destroyed by the frost; notwithstanding this severe check, it again recovered, and afterward became a fine tree, between twenty-five and thirty feet high, with a large, well-proportioned head, and a trunk of four feet in circumference, the lower branches sweeping the ground, and the whole tree producing annually from three hundred and fifty to four hundred large, elegant, and fragrant flowers. The seeds, however, never arrive at perfect maturity, although the fruit attains its full size and remains upon the tree till the following spring. This tree still exists, and is now upward of thirty feet high and more than one hundred years of age.

At Caserta, in the neighborhood of Naples, this tree has attained the height of nearly sixty feet. In this climate they also ripen seeds freely.

LONG-LEAVED MAGNOLIA.

MAGNOLIA MACROPHYLLA. MICHAUX, Flor. Bor. Am., vol. i. p. 327.
MICH., Sylva, pl. 57.

THE principal locality of this fine species has been for many years confined to the vicinity of Lincolnton, North Carolina, ten miles southeast of the town, near or on the estate of a man named Smith. The trees occupy the banks of a small stream, and are chiefly dispersed over its acclivities, in a rich loamy soil. In Tennessee, near the Cumberland River, I afterward saw a few small trees of this species; but in the winter of 1830, in an extensive tour which I made through the interior of the Southern States, I met with abundance of the *Magnolia macrophylla*, seventy miles from Tuscaloosa, on the banks of the Coosa in Bibb county, Alabama, growing often in the same bottom-lands as the *Illicium Floridanum*. To give me some idea of the magnitude of the Magnolias growing in this vicinity, a farmer told me some trunks produced sixteen *rails to the cut*, and that the trees were two feet in diameter; but as timber it was little esteemed, not enduring long in the air. To the town of Cahawba, I still saw the Magnolia; and afterward, in the immediate vicinity of Tuscaloosa, on the road down the banks of the Black Warrior, toward Florida, I observed this species in the greatest abundance, often as much as sixty feet in height; but, being the depth of winter, I of course could form no adequate conception of the splendor of its appearance when in vigorous vegetation.

EAR-LEAVED MAGNOLIA, (*Magnolia auriculata*, LAM. *Long-leaved Cucumber Tree*, MICH., Sylva, pl. 56.) In Bartram's Garden, at Kingsessing, in this vicinity, there is a tree of this species seventy or more feet high, and with a trunk of the diameter of two to three feet.

UMBRELLA TREE, (*Magnolia tripetala*.) According to Professor Torrey, no person since Michaux has found this tree in any part of the State of New York.

TULIP TREE, (*Liriodendron tulipifera*, LINN.) Respecting the northern limits of this tree, G. B. Emerson, Esq., informs me, "I have found a single tree of Liriodendron in Norfolk county, Massachusetts. It is plentiful on Westfield River."

DRIMOPHYLLUM.*

Natural Order, LAURINEÆ, (Ventenat.) *Linnæan Classification*.
ENNEANDRIA, MONOGYNIA.

Flowers HERMAPHRODITE.—*Perigonium* 6-cleft, subcampanulate, the segments spathulate-linear, equal, deciduous, the base persistent. *Stamina* nine, in three series, all fertile, filaments short, the three of the inner series with a pair of roundish, large, sessile glands, near the base of each. *Anthers* elliptic-ovate and similar, all opening on the inner side, 4-celled, the cells equal and parallel, with all the valves ascending. *Ovarium* 1-celled, with one ovule. *Style* short; *stigma* somewhat depressed, capitate. *Berry* 1-seeded.

A sempervirent, small tree of Upper California, with alternate, lanceolate, minutely reticulated, pungently aromatic, smooth leaves. Flowers small, yellow, smooth, in small terminal naked clusters. (Nearly allied to *Ocotea* of Aublet, but with hermaphrodite flowers, similar and uniformly introrse anthers, and a deciduous perianth. The leaves are also naked, without pennate nerves, and the inflorescence not in panicles. It is also nearly allied to *Aperiphracta* of Nees, but with flowers of a very different habit, and, with those genera, belongs to the tribe OREODAPHNEÆ.)

---

* From Δριμυς, *acrid to the taste*, and φυλλον, *a leaf*.

CALIFORNIAN BAY TREE.

DRIMOPHYLLUM PAUCIFLORUM.
OCOTEA SALICIFOLIA? KUNTH, SYNOPS., vol. i. p. 458.

THIS is a very elegant evergreen tree of Upper California, growing round Santa Barbara, twenty to twenty-five or thirty feet high, with erect, terete, and smooth branches. The wood is white and rather soft. The leaves are alternate, evergreen, coriaceous, perfectly smooth, three to four inches long and three-quarters to one and a quarter inches wide, lanceolate-pointed, but obtuse, entire, with very indistinct slender lateral nerves, and strongly but minutely reticulated above; the foot-stalks are about two to three lines long. The odor and taste of the leaves are very aromatic, the latter so much so as to be quite pungent, even more so than the leaves of the Bay; and they are employed as condiments by the inhabitants. The flowers are in small contracted clusters, at first surrounded with bud scales, which are caducous, but not in the form of an involucrum. The flowers are about four or five together, on pedicels nearly as long as themselves. The perianth is yellowish, funnel-formed, and somewhat spreading, deeply 6-cleft, the segments linear-spathulate and smooth, a little pubescent within toward the base. Stamens nine, with short and broad filaments, the anthers oblong, 4-celled, all opening from within, the cells parallel and nearly all equal, with the valves ascending, the three innermost each furnished toward the base of the filament with two large reniform, sessile glands. The perianth is deciduous, the base alone being persistent, and enlarging with the 1-seeded berry. The perfect fruit I have not seen.

PLATE XXII.

A branch of the natural size. a. The flower enlarged.

Drimophyllum pauciflorum.
Californian Bay tree   *Drimophylle pauciflore.*

UMBELLULARIA,

(NEES, ab Esenb.)

Natural Order, LAURINEÆ. *Linnæan Classification*, ENNEANDRIA, MONOGYNIA.

Flowers HERMAPHRODITE.—The *perianth* deeply 6-parted and shortly campanulate, the segments equal and deciduous. *Stamina* twelve, the nine exterior fertile, interior sterile, the three fertile innermost ones each with a pair of large glands covering the back of the filament. *Anthers* 4-celled, those of the third series with the two upper cells introrsely opening, the lower extrorsely and lateral; the sterile filaments thread-shaped. *Stigma* peltate, subrepand. *Berry* 1-seeded, seated on the cup-shaped base of the perianth.

A tree of Upper California, with alternate pennately-nerved leaves. The flowers aggregated in axillary umbellated clusters, enclosed by the involucrum; the scales of the involucrum broadish, alternate, and approximate; after the opening of the flower, caducous.

CALIFORNIAN UMBELLULARIA.

UMBELLULARIA CALIFORNICA. *Hermaphrodita, foliis perennantibus oblongo-lanceolatis vix acutis penninerviis reticulatu-venosis glabris, pedunculis axillaribus simplicibus, floribus pluribus, umbella subcapitata, pedunculo folio brevioribus.*

TETRANTHERA? CALIFORNICA, HOOKER et ARN., in Bot. Beech. Voy., p. 158. HOOKER, Flor. Bor. Am., vol. ii. p. 137.

LAURUS REGIA? DOUGLAS, Journal, in Hook. Compan. Bot. Mag., vol. ii.

THIS splendid evergreen tree was discovered, on the northwest coast, in Upper California, by Mr. Menzies, who first made known to botanists the vegetable treasures of that interesting and then unexplored region. Douglas afterward found it in nearly the same country, south of the Columbia or Oregon, and adds, that it attains the height of from forty to one hundred and twenty feet, with a diameter of from two to four feet. It commences at the southern limit of the prevailing Pine and Fir forests which line the wastes of Oregon. The foliage gives out, when bruised, a most powerful camphorated odor, which from its pungency is capable of exciting sneezing. Flowering specimens of this interesting tree were in Douglas's collection from California. It is to be regretted, however, that no detailed description nor figure is given; and I had not the good fortune to meet with it myself.

SASSAFRAS, (*Laurus sassafras*, LINN.) The inhabitants of North and South Carolina distinguish two kinds of Sassafras, the Red and the White. The Red or true *L. sassafras* I referred (in the "Genera of North American Plants," vol. i. pp. 259, 260,) to a sub-genus *Euosmus*, embracing also the following variety, which I then considered as a species, by the name of *L. (Euos-*

mus) *albida.* It is distinguishable from the Red by having the buds and twigs smooth and glaucous; its leaves are also smooth and thin, and the veins almost obsolete beneath; the petiole is apparently longer. The root is much more strongly camphorated than that of the Red sort, and is nearly white. It is better calculated to answer as a substitute for ochra (*Hibiscus esculentus*) than the common kind, as the buds and young branches are much more mucilaginous. It is abundant in North and South Carolina, from the Catawba Mountains to the east bank of the Santee, growing with the common kinds.

From the present order of plants we derive the Cinnamon, Cassia, and the Camphor. Several species afford the cinnamon of commerce, and the *Laurus Quixos* produces that of Peru. The cinnamon of Santa Fe de Bogota is afforded by *Laurus cinnamomoides.* A great deal of the finest camphor of India, however, is the product of the *Dryobalanops camphora.* The volatile oil obtained from some species of Laurus found in the vast forests between the Orinoco and the Parime is produced in great abundance by merely making an incision into the bark with an axe, as deep as the liber or young wood. It gushes out in such quantities that several quarts may be obtained by a single incision. It has the reputation of being a powerful discutient.

THE LINDEN, or LIME TREE.

Natural Order, TILIACEÆ, (Jussieu.) *Linnæan Classification*,
POLYANDRIA, MONOGYNIA.

TILIA.* (LINN.)

Sepals five. *Petals* five. *Stamens* numerous, disposed more or less in five clusters, the central tuft (chiefly in the native species) transformed into a petal. The *ovary* globular, villous, and 5-celled, each of the cells bearing two ovules. *Capsule* ligneous, globular, by abortion only 1-celled, with one or two seeds. *Cotyledons* sinuate.

Trees of Europe and North America, with alternate dilated or cordate leaves, oblique at the base, serrated on the margin, and with a tough and fibrous bark; stipules caducous. The flowers disposed in flattish pedunculated clusters, (or cymes,) and with the peduncle curiously adnate for a great part of its length to a large membranaceous, linear bracte. The rest of this family of plants are nearly all tropical productions.

---

* An ancient Latin name, probably from the Greek πτελεα, the ELM.

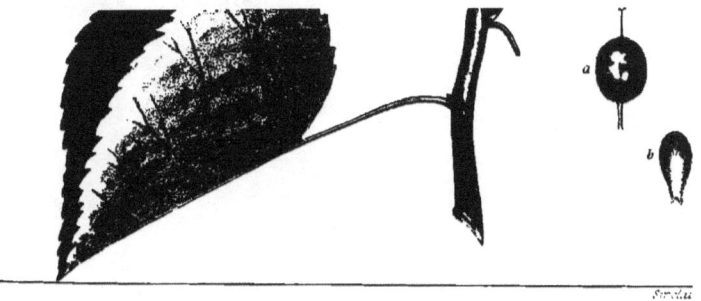

Tilia heterophylla.

Large-leaved Linden

Tilleul heterophylle

LARGE-LEAVED LINDEN, or LIME.

TILIA HETEROPHYLLA. *Foliis ovatis, argute serratis, basi nunc cordatis, nunc oblique aut æqualiter truncatis, subtus tomentosis; nuce pisiformi.—* VENTENAT, Mem. de l'Institut., tom. 4, p. 16, pl. 5. PURSH, Flor. Bor. Am., vol. ii. p. 363. NOUVEAU DUHAMEL, vol. i. p. 229. DECANDOLLE, Prod., vol. i. p. 513.

TILIA HETEROPHYLLA. Leaves glabrous and deep green above, very white and velvety-tomentose beneath, the veins dark-colored and nearly glabrous, with coarse mucronate serratures; petals obtuse, crenulate; staminodia (inner petals) spatulate, entire, style hairy at the base.—TORREY and GRAY, Flor. North Amer., vol. i. p. 239.

TILIA ALBA.—*Smith's* Insects of Georgia, vol. i. p. 21, t. 11?

This is one of the rarest and most ornamental trees of the whole genus; and, as far as my own observations go, it is almost wholly confined to the shady forests of the Ohio and its tributary streams, to which Pursh also adds the banks of the Mississippi. Torrey and Gray received it likewise from the neighborhood of Macon, in Georgia, where it was collected by our late mutual friend and excellent observer, Dr. Loomis. In descending the Ohio, late in autumn, (about the year 1816,) I got out of the boat in which I was descending, to walk round Le Tart's Rapids above Cincinnati. Here I observed almost an exclusive forest of this fine Linden, on a rather-elevated alluvial platform, in a light, rich, calcareous soil. Most of the trees were tall and rather slender, sixty to eighty feet in height, and the ground was thickly strewed with their large and singular leaves, almost as white as snow beneath. According to the herbarium of Mr. Schweinitz, it exists also in Virginia, probably on the borders of the streams which flow into the Ohio near Pittsburg; and according to Dr. Short, of Lexington, Kentucky, it forms in his vicinity one of the largest forest trees in the rich lands there. Decandolle speaks of having received a specimen of some very similar species from Mexico. It does not yet appear to have been introduced into Europe, though it is properly described in the New Duhamel, probably from Ventenat's essay, as the leaves are said to be snow-white beneath.

The young branches are purplish and somewhat glaucous. The largest leaves I have seen are about six or seven inches long and three to five broad. In the young state, the white pubescence beneath is most conspicuous when the leaves are thinly covered; the hairs are stellate, the serratures are strong and sharp, with acuminated rigid points; the upper surface is dark green: the base of the leaf varies considerably; sometimes it is sinuated, at other times perfectly flat and truncated; the leaves are always very oblique at the base. The flowers are somewhat larger than those of *T. Americana*, and the fruit is villous, nearly spherical, and certainly always without any ribs.

The TILIA ALBA, White Lime of Michaux, plate 132, not being the *T. alba* of Kitaibel and Aiton, (Hort. Kew. l. c.,) which is a native of Hungary, it is necessary to change its name, and we propose to call it TILIA MICHAUXII, (Michaux's Lime,) if his plant should indeed prove to be any thing more than a smoother variety of our *T. heterophylla*.

PLATE XXIII.

A branch of the natural size. a. *The fruit.* b. *The flower.*

General Observations. The Lime has long been a favorite tree for avenues and public walks; it is planted in the streets of some of the principal towns of France, Holland, and Germany, and it is used for forming avenues both on the continent of Europe and in Great Britain. It has of late years been much planted along the streets in several towns and cities of the United States, but in Philadelphia it is so much attacked by insects that it probably will not long survive. The species employed for this purpose appears to be principally the European, while the native kinds, being more hardy and vigorous, ought to have the preference, particularly the present species, (*T. heterophylla,*) which in a good soil becomes a large tree, and is at the same time splendidly ornamental. The insect that devours the leaves of the Linden appears to be a moth,[*] which suspends its cocoons at the ends of the twigs of the trees it has stripped; these ought carefully to be removed and destroyed, by which means the evil, if not wholly cured, would be decidedly mitigated.

The Dutch plant the Lime in towns, along their widest streets, and by the sides of their canals; and the whole country is thus perfumed by their flowers during the months of July and August: they likewise afford an ample repast for the bees.

The wood of the European Lime Tree is of a pale yellow or white, close-grained, soft, light, and smooth, and not liable to be attacked by insects. It is used by pianoforte-makers for sounding-boards, and by cabinet-makers for a variety of purposes. It is turned into domestic utensils of various kinds, carved into toys, &c. The most elegant use to which it has been applied is for carving, for which it is superior to any other wood. Many of the fine carvings in Windsor Castle, St. Paul's, Trinity College Library at Cambridge, and in the Duke of Devonshire's mansion at Chatsworth, from the hand of the celebrated Gibbons, are of this wood. It makes excellent charcoal for gunpowder. Baskets and cradles were formerly made from the twigs. The leaves are also employed as fodder for cattle in Europe. It is in Russia and some parts of Sweden

---

[*] A species of *Oiketicus*.

that the well-known bass mats are formed from the inner bark of this tree. The bark stripped from young trees of six inches to a foot in diameter is selected for this purpose. These strips are steeped in water till the bark separates freely into layers; it is then taken out and separated into strands, which are dried in the shade, and afterward manufactured into the mats so much used by gardeners and upholsterers, and for covering packages. The fishermen of Sweden make fishing-nets of the fibres of the inner bark, formed into a kind of flax; and the shepherds of Carniola even weave a coarse cloth of it, which serves them for their ordinary clothing. The whole plant abounds with mucilage, the sap, like that of the Maple, affords a considerable quantity of sugar, and the honey produced by the flowers is considered superior to all other kinds for its delicacy, selling at three or four times the price of common honey; in Europe, it is used exclusively in medicine, and for making some particular kinds of *liqueurs*, especially rosolio. This Lime Tree honey is only to be procured at the little town of Kowno, on the river Niemen, in Lithuania, which is surrounded by an extensive forest of Lime Trees. The triturated fruit produces also a paste very similar to that of cocoa. During the taste for grotesque decorations, the Lime, like the Yew, was cut into various imitative forms, and in some of the public gardens of recreation round Paris and Amsterdam there are very imposing colonnades, arcades, walls, pyramids, and other architectural-looking masses formed of this tree.

The European Linden attains a height of upward of one hundred feet, and grows with vigor for several centuries. In Switzerland there are some very large and ancient Lime Trees: one, mentioned by Decandolle the younger, near Morges, has a trunk of twenty-four feet four inches in circumference; another, near the great church at Berne, which was planted before the year 1410, is thirty-six feet in girth.

MANGLE.

Natural Order, RHIZOPHOREÆ, (R. Brown.) *Linnæan Classification*, DODECANDRIA, MONOGY.

RHIZOPHORA.* (LINN.)

Tube of the *calyx* obovate, coherent with the ovary, the border divided into four oblong, persistent segments. *Petals* four, oblong, emarginate, coriaceous, conduplicate, before expansion embracing the alternate stamens, the margins each with a double row of long, woolly hairs. *Stamens* twice as many as the petals; *anthers* nearly sessile, large, linear-oblong. *Ovary* 2-celled, with two ovules in each cell. *Style* conical, short, 2-furrowed; *stigma* 2-toothed or bifid. *Fruit* ovate or oblong, crowned near the base with the persistent segments of the calyx, longer than the tube, at length perforated at the apex by the radicle of the germinating embryo.

Maritime trees of the tropics, with entire opposite leaves and axillary flowers.

---

* The name, from ριζα, a *root*, φερω, to *bear*, in allusion to the seed germinating before it falls from the branches.

AMERICAN MANGLE, OR MANGROVE.

RHIZOPHORA AMERICANA. *Foliis obovato-oblongis obtusis; pedunculis trichotomis petiolo longioribus, stylis subulatis bifidis, fructibus subulato-clavatis obtusis.*
RHIZOPHORA *mangle.*—JACQUIN, Amer., p. 141, t. 89. BROWN, Jam., p. 211. DECAND., Prod., vol. iii. p. 32. NUTT., Florid., pl. Sill. 5, p. 295. TOR. and GRAY, vol. i. p. 484, (not of LINN.)
Candela Americana foliis laurinis.—CATESBY'S Carol., vol. ii. p. 63, t. 63.
Mangle aquatica, foliis subrotundis et punctatis.—PLUMIER, Gen., p. 13. SLOANE, Jam., p. 155, Hist., vol. ii. p. 63.
Margue Guapariba.—PISO's Brazil, l. 4, c. 87, R.

THIS tree is found in the maritime swamps of Louisiana and East Florida, and along the coast of Texas is not uncommon. The Mangrove, like the famous Banyan Fig, sends out innumerable roots into the surrounding marshes from the fusiform fruits which terminate its branches, so that after a while a single tree becomes, as it were, the parent of a whole forest of several miles in extent; and, growing well even into the salt water, it is not unfrequent to see their branches loaded with oysters (the *Ostrea folium*) of an exquisite flavor. Those thickets likewise afford a resort for various kinds of sea-fowl, and, fringing the margin of the ocean and the salt-pools with their spreading summits, they give a peculiar feature to the tropical landscape, but at the same time afford shelter to clouds of mosquitos. The bark and fruit are useful for tanning: the flower, according to Loureiro, dyes a very durable black, and, according to Sloane, affords a material for ink.

The Mangrove of the West Indies and Tropical America becomes a tree about forty to fifty feet high and two to three feet in diameter, with a ferruginous bark and white wood of no great value except for fuel; yet, according to Sloane, the wood

Rhizophora Americana.
American Mangle

is good for building and for shingles. The wood of that of India, as described by Roxburgh in his "Flora Indica," is of a dark-reddish color, hard, and durable.

The Mangrove is not very tall, but very branching; the branches, almost always opposite, elongated and pendant. When touching the soil, they strike root and become new trees, which remultiply themselves in the same manner, thus forming an almost impenetrable barrier on the borders of the sea.

The leaves are opposite, entire, coriaceous, at first folded inward, with caducous stipules between the petioles. The flowers are pale yellow, the segments of the calyx lanceolate. The anthers are subulate; the margin of the petals pilose; the style bifid, with the divisions rather long and subulate. The verdure of the Mangrove is dark and gloomy, and the whole tree, inhabiting a region of desolation, presents an aspect of sadness.

The most extraordinary plant of this, or rather a nearly-allied genus, is the *Rhizophora gymnorhiza* of Linnæus, (now *Brugiera*.) This tree grows commonly in the maritime marshes of India; and the branches of its numerous roots, ascending into the air, produce the appearance of a large umbrageous tree, as it were, on stilts, or, as Roxburgh says, supported in the air on a circle of converging hop-poles. The fruit, the leaves, and the bark of this species are also said to afford food to the native inhabitants. A figure of it is given by Rumphius, vol. iii. t. 68, and by Rheede, in the "Flora Malabarica," vol. vi. tab. 31, 32.

PLATE XXIV.

A branch of the natural size. a. *The fruit.*

GUAVA.

Natural Order, MYRTACEÆ. *Linnæan Classification*, ICOSANDRIA, MONOGYNIA.

PSIDIUM.* (LINN.)

Calyx-tube (or external germ) ellipsoid or obovate, often contracted at the summit; the border at first undivided and ovate while in flower, afterward 1 to 5-cleft. *Petals*, five. *Stamens* very numerous, distinct. *Style* filiform; *stigma* capitate. The *ovary* with from five to twenty cells, some of them abortive, each cell subdivided by the interposition of a placenta resembling a dissepiment. *Ovules* numerous, horizontal. *Fruit* a many-seeded berry, coated with the adhering tube of the calyx and crowned by its persisting lobes. The *seeds* scattered through the pulp in the ripe berry, having a bony or hard shell. The *embryo* curved in a half-circle round the protruded base of the testa. *Cotyledons* minute; the radicle rather long.

Trees or shrubs chiefly indigenous to the intertropical regions of America, with opposite, entire, impunctate, feather-nerved leaves. Peduncles axillary, 1 to 3-flowered, each flower with a pair of bractes. The flowers white.

---

* One of the Greek names for the Pomegranate. *Guava* is a corruption of the American aboriginal name of *Guayaba*.

Pisidium buxifolium.

Florida Guara.   Goyavier de la Floride.

FLORIDA GUAVA.

PSIDIUM BUXIFOLIUM. *Glabrum, ramulis teretibus, foliis parvulis coriaceis cuneato-obovatis obtusis subsessilibus margine revolutis, pedunculis solitariis brevissimis unifloris, fructu pyriformi.*

For a knowledge of this interesting tree or shrub we are indebted to the late indefatigable Dr. Baldwin, who met with it in some part of East Florida near the river St. Johns. To show how very unlike this species is to all the others known, it was hastily marked by Mr. Schweinitz, in his herbarium, (of which the specimen forms a part,) "*Quercus virens,*" and at the first hasty glance some resemblance may be traced with the Live Oak in the leaf and twig; but, of course, the presence of the fruit at once dispels the illusion.

I have seen but the single specimen now figured, and would recommend its examination to some future traveller. The twig is round, covered with a gray bark, and at near distances marked with the cicatrices of opposite fallen leaves. The leaves on the upper branchlets are crowded together in opposite pairs, of a very thick, opaque, rigid consistence, and appear to be sempervirent; they are perfectly smooth on both sides, paler beneath, dark-green above, cuneate-obovate, obtuse, sometimes with an attempt at a very short and blunt acumination, with the margin reflected, and beneath marked with numerous approximating feathered nerves; they are from one inch to one and a half inches long by one-half to three-quarters of an inch wide. The peduncles are axillary and solitary, very thick in the fruit-stalk, and scarcely two lines long. The flowers I have not seen. The berry is blackish-purple, pear-shaped, about the size of a cherry, and appears to have been succulent, as usual; internally it is filled with horizontal rows of flat, subreniform, pale-brownish, bony seeds, with a narrow embryo curved into the form of a horseshoe. The cotyledons are very small, and in the seed are of a bright waxy yellow. This species is very nearly allied to the Purple-fruited Guava, (*P. Cattleianum,*) scarcely differing in any thing but the smallness of the leaves and the pyriform fruit, though the leaves of the Purple Guava, besides being much larger, are also pubescent when young. Most of the species of this genus are cultivated in the tropics for their fruit. The *P. pyriferum*, or Common Guava, bears a fruit about the size of a hen's egg, yellowish, with a peculiar odor; the pulp is rather firm, flesh-colored, sweet, agreeable, and aromatic. In the West Indies it is highly esteemed by all classes, being eaten raw, as a dessert, or formed into an excellent sweetmeat and jelly.

Of the fruit of the Purple Guava, to which ours is so closely related, Lindley remarks, "The excellent flavor of its fruit, which is very like that of strawberries and cream, is far superior to either *P. pyriferum*, *pomiferum*, or *polycarpon*." Mr. Sabine remarks of the fruit of this species, "that it is juicy, of a consistence much like that of a strawberry, to which it bears some resemblance in flavor."

What the present species may become, when cultivated, remains to be proved; but in a genus so generally interesting for their fruit, the experiment is worth making when an opportunity may offer. Probably Dr. Baldwin found it growing near or above New Smyrna, as he did not go much farther into the interior of East Florida.

PLATE XXV.

A branch of the natural size in fruit.

Calyptranthes chytraculia.

Forked Calyptranthes. *Calyptranthe chytraculie*

CALYPTRANTHES.*

(SWARTZ.)

Natural Order, MYRTACEÆ. *Linnæan Classification,* ICOSANDRIA, MONOGYNIA.

Tube of the *calyx* obovate, with the border entire; when flowering, bursting circularly in the form of a lateral, and at length deciduous, lid. *Petals* none, or two or three and minute. *Stamens* many. *Style* one; *stigma* simple. *Ovary* 2 or 3-celled, the cells 2-seeded. The *berry* by abortion 1-celled, 1 to 4-seeded.

Small trees of the West India Islands and of Brazil, the leaves with pinnated veins. Flowers small and numerous, usually in axillary or terminal panicles.

FORKED CALYPTRANTHES.

CALYPTRANTHES CHYTRACULIA. *Arborea, foliis ovatis apice attenuatis rigidiusculis demum glabris, pedunculis axillari-terminalibus trichotomis paniculatis floribusque rufo-velutinis.*—DECAND., Prod., vol. iii. p. 257.
C. CHYTRACULIA. *Arborea, pedunculis terminalibus trichotomis tomentosis, foliis ovatis apice attenuatis.*—SWARTZ, Prod., p. 79; Flor. Ind. Occid., vol. ii. p. 921.

* The name from καλυπτρα, *a veil,* and ανθος, *a flower,* in allusion to the operculid form of the calyx.

MYRTUS CHYTRACULIA.   *Pedunculis dichotomis paniculatis tomentosis, foliis geminis subovatis terminalibus.*—LINN., Amœn. Academ., vol. v. p. 398. SWARTZ, Observ., p. 202.
Chytraculia arborea, foliis ovatis glabris oppositis, racemis terminalibus. —BROWN, Jamaic., p. 239, t. 37, fig. 2.
EUGENIA PALLENS? POIRET, Suppl., vol. iii. p. 122.

THIS plant forms an elegant and curious small tree, with hard wood, and in Jamaica is accounted an excellent timber; but the trunks seldom exceed fourteen or fifteen inches in diameter. In Jamaica it is found in the dry mountain-lands; it is also indigenous to the islands of St. Thomas and Guadaloupe, and it has now also been found on Key West by Dr. Blodgett.

The branches appear to be covered with a gray and smooth bark. The leaves, when in bud, as well as the young branches, flower-stalks, and calyx, are clad with a short, soft, ferruginous down, which wholly disappears from the leaves as they advance in their development; they are of a lanceolate-ovate form, narrowed into a short petiole below; above, acuminate but obtuse; beneath they are distinctly pennate-nerved, and too opaque to admit the light through the resinous glands with which they are nevertheless provided: they are about two inches long by an inch in width. The flowering panicles are trichotomous, usually terminal, and considerably ramified. The flowers are small and whitish, from the color of the stamens. The calyx is ferruginous and tomentose, formed of a small obovate even cup; the whole border, separating in a circular manner, flies over to one side, in the form of a rounded petal, from whence issue the numerous filiform stamens with small whitish anthers. The germinal fruit appears small, dry, and tomentose; but I am unacquainted with it in a ripe state.

PLATE XXVI.

A branch of the natural size. a. A flower magnified, showing the lateral adherence of the lid of the calyx.

EUGENIA.*

(MICHELI, LINN.)

Natural Order, MYRTACEÆ.   *Linnæan Classification*, ICOSANDRIA,
MONOGYNIA.

The tube of the *calyx* roundish, with the border deeply 4-parted. *Petals* four. *Stamens* many, free. *Ovary* 2 to 3-celled, the cells containing several ovules. *Berry* sub-globose, crowned with the persisting calyx; when mature, 1 or rarely 2-celled. *Seeds* one or two, roundish and large. The *embryo* pseudo-monocotyledonous, the *cotyledons* very thick and wholly blended together, the *radicle* more or less distinct and very short.

These are trees or shrubs mostly indigenous to the Caribbean Islands, or the warmer parts of America. The leaves and inflorescence are very similar to those of the Myrtles.

---

* So named in honor of Prince Eugene of Savoy, who was a protecter and encourager of botany, and possessed a botanic garden.

SMALL-LEAVED EUGENIA.

EUGENIA DICHOTOMA. *Pedunculis axillaribus oppositis et subterminalibus folio longioribus bifidis aut bis bifidis, floribus in dichotomiis sessilibus cæteris pedicellatis, foliis elliptico-lanceolatis basi attenuatis pellucido-punctatis, adultis glabris, junioribus utrinque ramulis calycibusque pubescentibus.*—DECAND., Prod., vol. iii. p. 278.
MYRTUS *dichotoma*.—Vahl! MSS., POIRET, Supplem., vol. iv. p. 53?
β FRAGRANS, *foliis ovatis glabris.* EUGENIA FRAGRANS. WILLD., Sp. Pl., vol. ii. p. 964. Bot. Magaz., t. 1242. *E. montana*, AUBL., Guian., vol. i. p. 495, t. 195?
EUGENIA *divaricata*, LAM., Encyc., p. 202.

This elegant and fragrant species of Eugenia, resembling a Myrtle, becomes, at Key West, according to Dr. Blodgett, a tree. It is also indigenous to the islands of St. Domingo and Cuba, where specimens have been collected by Poiteau and La Sagra. The variety *fragrans* (for such I must consider it) is a native of the high mountains in the southern part of Jamaica and Martinique, and, if the same with Aublet's *E. montana*, is also a native of Guiana. The *E. fragrans* has many years since been collected by Dr. Baldwin, in the vicinity of New Smyrna, in East Florida.

The wood of *E. divaricata*, according to Lamarck, is hard, close-grained, and reddish, and is much esteemed for articles of furniture. The wood of the Florida Tree is exactly similar; while that of *E. montana*, according to Aublet, is hard, compact, and white.

The branches of the plant now figured are covered with a smooth, light-gray or silvery bark, and at the summits are crowded with small, shining, almost opaque, leaves, but yet interspersed with the usual resinous vesicles of the genus; they are from an inch to an inch and a half in length, and about three-quarters of an inch in breadth, mostly elliptic or elliptic-

Eugenia dichotoma.

Small leaved Eugenia. Tambusier dichotome

SMALL-LEAVED EUGENIA.

oblong, and always narrowed below; sometimes they are nearly lanceolate and obtuse at the point; scarcely any veins are visible on either side, but the midrib is prominent beneath. The young leaves, buds, peduncles, and calyx are clothed with a close, short, hoary pubescence, which in the variety *fragrans* is much less distinct or almost wanting. The peduncles are axillary, coming out toward the summits of the branches, and are of various lengths, sometimes only a little longer than the leaves, at other times crowded into trichotomous branchlets two or three times longer than the leaves; in their most simple form, except by the abortion of the lateral buds, they terminate in three flowers, the central one sessile in the fork, and the lateral ones on longish, diverging pedicels; at other times the peduncles are twice trifid, or even more ramified, and lengthened out very much in the progressive ripening of the fruit. The segments of the calyx are always four, broad and rounded, covered with resinous cists or vesicles, and pubescent or ciliate on the margins. The petals are likewise rounded or concave, whitish, with a tinge of red. The stamens are numerous. Style simple and subulate. The berry at length only 1-seeded. There are a pair of minute, subulate bractes under the base of each flower-bud, but so deciduous that they are seldom to be seen.

One of the specimens of the variety *fragrans*, from New Smyrna, has very slender twigs; and on the same specimen there are obtuse and very *sharply-acute* leaves. In this also the peduncles are chiefly axillary. This plant is nearly as fragrant as the common Myrtle.

PLATE XXVII.

A branch of the natural size.   a. The flower a little enlarged.   b. The berry, of the natural size.

TALL EUGENIA.

EUGENIA PROCERA. *Pedicellis unifloris axillaribus 2–4-confertis folio brevioribus sub flore bibracteolatis, foliis ovatis obtuse acuminatis ramisque glabris.*—POIRET, Suppl. Encyc., vol. ii. p. 129. DECAND., Prod., vol. iii. p. 268.

MYRTUS PROCERA. *Pedunculis confertis axillaribus unifloris, foliis ovatis acuminatis planis glabris, ramis virgatis, caule arboreo.*—SWARTZ, Prod., p. 77. Flor. Ind. Occident., vol. ii. p. 887. WILLD., Sp. pl., vol. iv. p. 968.

THIS is another plant with the aspect of a Myrtle, which becomes a tree and attains an elevation of twenty to thirty feet. It was discovered by Swartz in the forests of the interior of Hispaniola. It is likewise indigenous to the islands of Martinique and Santa Cruz, and has now been found common on Key West by the same gentleman who met with the preceding species.

The wood appears to be white and close-grained. The twigs are clothed with a light gray, almost white and silvery, bark, and are spreading and sometimes zigzag. The leaves are on short petioles one and a half to two and a half inches long by an inch to an inch and a half wide, ovate-acuminate, and obtuse, rather opaque, nearly scentless, though provided with the usual resinous vesicles, and from the bud they are perfectly smooth. The flowers are said to be fragrant, and come out on separate axillary peduncles, from two to four together; the peduncles at first are not more than three or four lines long, but grow out at length to the extent of half an inch. The segments of the calyx are four, rounded and broad, rough, with aromatic vesicles, but smooth. The petals, four, are rounded and concave, slightly ciliated, and appear to have been reddish white. Stamens numerous, the anthers whitish. The berry spherical, brownish yellow, about the size of a grain of black

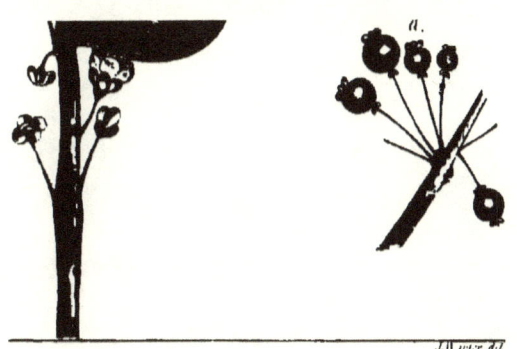

Eugenia procera.
Jambosier élevé

Eugenia buxifolia

Box leaved Eugenia.   Jambosier à feuilles de buis.

pepper, studded over with numerous glands or aromatic cists, and crowned with the broad, persisting border of the calyx; at first 2-celled, with several ovules; at length the berry is only 1-seeded; the seed large, with no distinct cotyledons. It flowers in April. The size of the leaves appears to vary, so that in some specimens they are uniformly only about half the dimensions we have given.

PLATE XXVIII.

A branch of the natural size and of the large-leaved kind. a. A cluster of the berries.

BOX-LEAVED EUGENIA.

EUGENIA BUXIFOLIA. *Pedunculis axillaribus ramosis multifloris brevissimis, pedicellis sub flore bibracteolatis, foliis obovato-oblongis obtusis basi attenuatis opacis subtus punctatis margine subrevolutis.*—DECAND., Prod., vol. iii. p. 275. WILLD., Sp. pl., vol. ii. p. 960.

MYRTUS BUXIFOLIA. *Racemulis brevissimis confertis axillaribus, foliis cuneatis oblongis obtusis convexiusculis.*—SWARTZ, Prod., p. 78. Flor. Ind. Occid., vol ii. p. 899. *M. monticola?* SWARTZ, Flor. Ind. Occid., vol. ii. p. 898.

MYRTUS AXILLARIS. POIRET, Dict., vol. iv. p. 412, (non Swartz.) *M. Poireti*, Spreng. Syst., vol. ii. p. 483.

THIS plant, also a native of Cuba, St. Domingo, and Jamaica, has been observed at Key West by Dr. Blodgett, where it is very common in sterile places, affecting the vicinity of the sea, and becoming a tree of about twenty feet in height, with a hard, white, close-grained wood. The bark is whitish-gray and even; the twigs are slender, and chiefly clothed with leaves toward their summits; they are wedge-oblong, sometimes almost lanceolate, obtuse, and always narrowed below into a minute petiole, so that they appear to be nearly sessile, above of a darkish green and somewhat shining, beneath dull and paler, slenderly nerved beneath, somewhat opaque, punctate, and slightly revolute on the margin; they are about one and a half inches long by one-half to three-quarters of an inch wide. The flowers are very small, in axillary branching clusters of three to seven together on the minute and very short bracteate raceme; there are two minute bracteoles under each flower; the calyx as well as the petals are studded with resinous glands, and the latter are more than twice the length of the calyx. The calyx, racemes, and minute branchlets are covered with a close brownish pubescence. The flowers are *polygamous*, on many specimens sterile, though furnished with the pistillum; and many of the flowering clusters are produced on the naked branches where they have been preceded by the former leaves. The berry is dark brown, covered with resinous glands or cists, about the size of a grain of black pepper, and when mature contains one or more (rarely two) large seeds in one or two cells, with blended, inseparable cotyledons.

PLATE XXIX.

A branch of the natural size.   a. A flower enlarged.   b. The berry.   c. A berry with two cells.

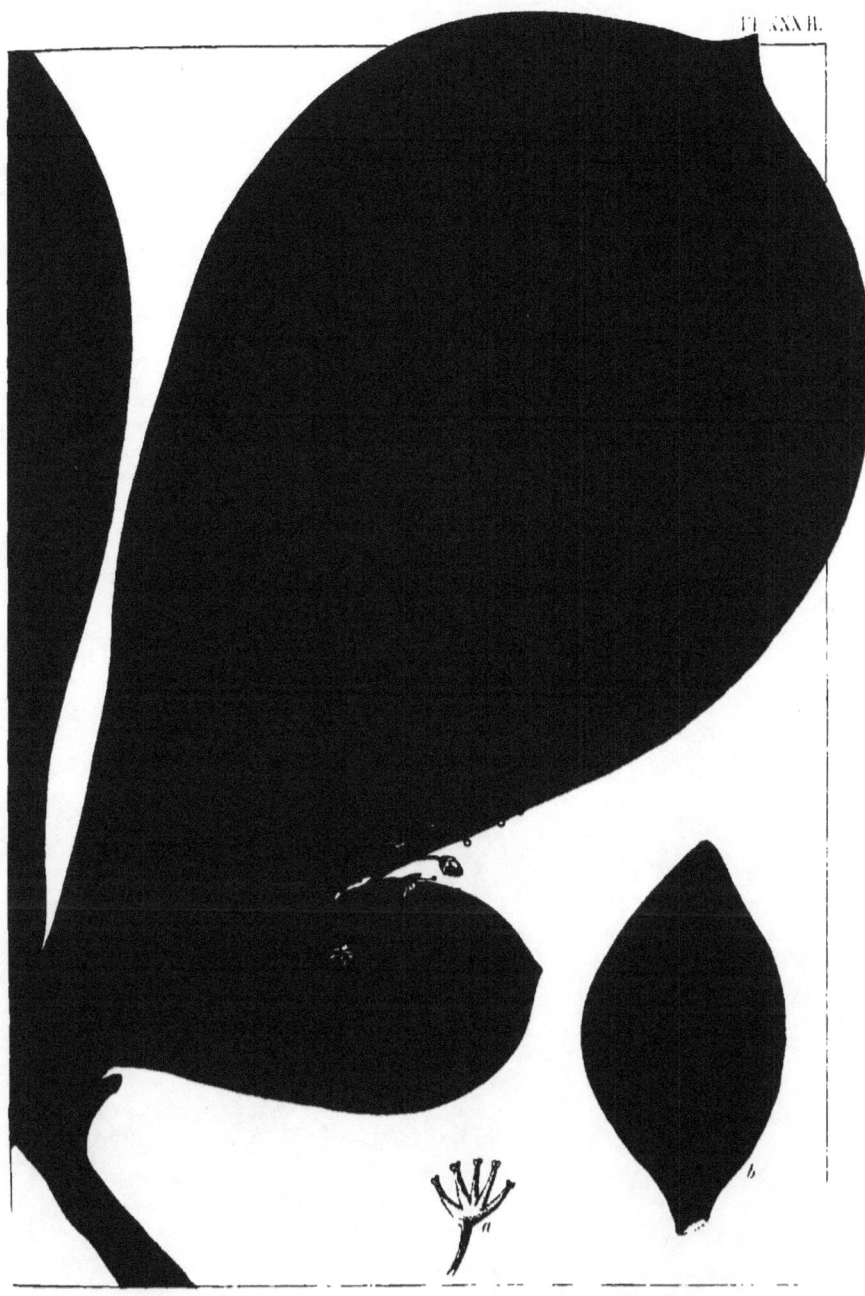

Terminalia catappa.

Indian Almond. Badamier de Malabar.

INDIAN ALMOND.

Natural Order, COMBRETACEÆ, (R. Brown.) *Linnæan Classification*, DECANDRIA, MONOGYNIA.

TERMINALIA. (LINN. DECAND.)

Flowers often POLYGAMOUS from abortion.—Border of the *calyx* deciduous, campanulate, 5-cleft, the divisions acute. *Petals* none. *Stamens* ten, in a double row, longer than the calyx. *Ovary* with two or three ovules. *Style* filiform, somewhat acute. *Drupe* not crowned by the calyx, often dry, indehiscent, 1-seeded. The *seed* resembling an almond. *Cotyledons* spirally convolute.

Trees of the largest size or shrubs, with alternate or rarely opposite leaves, crowded toward the extremities of the branches, and hence the generic name. Flowers in spikes; the spikes in racemes or panicles, bisexual in the lower part, and male in the upper.

§ I. CATAPPA, (Gærtner.) *The drupe compressed, with the margin winged or much attenuated.*

CATAPPA, OR INDIAN ALMOND.

TERMINALIA CATAPPA. *Foliis obovatis basi attenuatis subtus molliter pubescentibus, glandulis minimis subtus in basi folii ad latus nervi medii.*—DECAND., Prod., vol. iii. p. 11. LINN., Mantis., p. 519.

TERMINALIA CATAPPA. Leaves about the extremities of the branchlets on short petioles, obovate, cuneate, and attenuated, at the same time slightly cordate at the base, a little repand, with a large depressed gland beneath on each side the midrib near the base; racemes axillary, solitary, simple, shorter than the leaves; drupe oval, compressed, glabrous, with elevated navicular margins, convex on both sides.—ARNOT, Prod. Ind. Orient., vol. i. p. 313. JACQUIN'S Ic. Rar., vol. i. tab. 197. LAM., Illust. tab. 848, fig. 1. *Adamarum*, RHEED, Flora Malabarica, vol. iv. tabs. 3 and 4. TORREY and GRAY, Flor. N. Amer., vol. i. p. 485.

ACCORDING to Torrey and Gray, Dr. Hasler has discovered this splendid tree in Southern Florida. A variety of it is known to exist in the Caribbean Islands, which Humboldt and Kunth imagined to be introduced; but for this supposition there is probably no sufficient ground, as Poiteau collected it in the forests of St. Domingo, of which I have a specimen now before me. A near congener, if not the same thing, was found in Guiana by Aublet,—his *Tanibouca;* yet the favorite region of its existence is in the tropical forests of India, on the sandy and gravelly coasts of Malabar, and in the island of Java; it there becomes, according to Rheed, a very large and splendid tree of a pyramidal form, like that of a lofty Spruce, the leafy summit being composed of almost-horizontal branches disposed in circular stages. Its wood is white, very hard, covered with a smooth gray bark which is red within. The leaves, situated near to the extremities of the branchlets, six or seven together, at intervals, form circular clusters of great regularity; they are about six to nine inches long by three to five wide, of an inversely-ovoid or cuneate-oval figure, widening toward the summit, where they become almost round, with a short, abrupt, slanting point in the centre, narrowed and somewhat cordate at the base, nearly entire, or obscurely though sometimes very distinctly crenulated on the border, green and smooth above, slightly pubescent beneath; the young leaves and shoots, as well as the petioles, clothed with a brown and close tomentum. The flowers are small, without scent, of a whitish green, and disposed in great numbers in several almost terminal axillary slender spikes; they

CATAPPA, OR INDIAN ALMOND. 127

are nearly sessile, with caducous, concave, oval, pointed bractes. The calyx contains a small, very hairy, 5-toothed cup. The spikes are not as long as the leaves. The fruit is an elliptic shell, a little compressed, glabrous, surrounded with an elevated margin, convex on both sides, and reddish brown when mature. This dry drupe includes an oblong, very hard nut, of one cell, containing a white kernel, of a taste approaching to that of the filbert-nut, but more oily and soluble.

In India it is also cultivated in gardens. The large almond-like kernels of its nuts are eaten and served at the best tables. An oil is obtained from the kernels by expression, similar to that of the olive, which is said never to become rancid. It is made also into emulsions like almonds. The Indians employ the leaves medicinally for indigestion, bilious affections, and other maladies.

PLATE XXXII.

A small branch of the natural size. a. The flower. b. The nut.

Terminalia Benzoin has a milky sap, and was believed to produce the Benzoic acid, which, however, is now doubted.

Another of the species, *Terminalia vernix*, is said to afford the celebrated Chinese and Japanese varnish used in their lacquerware. This tree grows on the mountains of several of the southern provinces of China, and in the Moluccas. It possesses a lactescent juice, which, as well as its exhalations even, are said to be deleterious; but the kernels of its fruits, like those of the Catappa, are perfectly harmless and agreeable. At Batavia, regular plantations are made of the *Terminalia Moluccana*, in the gardens and places of public resort, for the sake of its agreeable shade.

CONOCARPUS.*

(GÆRTNER.)

Natural Order, COMBRETACEÆ. *Linnæan Classification*, PENTANDRIA, MONOGYNIA.

The flowers densely aggregated in globular or oblong spike-like aments.—Tube of the *calyx* about the length of the ovary, persistent; the border 5-cleft. *Petals* none. *Stamens* five to ten, exserted; the anthers heart-shaped. *Ovary* compressed, containing two ovules. The *fruits* coriaceous, corky, and scale-like, closely imbricated, and indehiscent. *Cotyledons* spirally convolute.

Small maritime trees or shrubs, with alternate, entire, somewhat coriaceous leaves. Heads of flowers pedunculated, axillary, or terminal, solitary or in panicles.

BUTTON TREE.

CONOCARPUS ERECTA. *Foliis oblongis utrinque acuminatis sæpius basi biglandulosis, capitulis paniculatis.*—DECAND., Prod., vol. ii. p. 16. JACQ., Amer., p. 78, t. 52. CATESBY'S Carolina, t. 33.
α *arborea*.—DECAND., l. c.
Conocarpus erecta. Fruits retrorsely imbricated in a subglobose head, somewhat boat-shaped, scarcely winged; tube of the calyx not produced beyond the ovary; leaves oval-lanceolate, mostly acute or acuminate at each end, usually with two glands at the base; heads panicled.—TORREY and GRAY, Flor. N. Amer., vol. i. p. 485.

---

* From κωνος, *a cone*, and καρπος, *a fruit*, its fruit resembling the cone of an Alder.

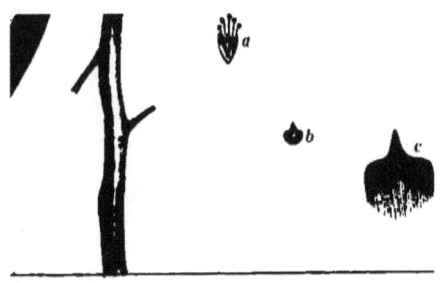

Conocarpus erecta.
Conocarpe droit.

BUTTON TREE. 129

Manghala arbor Curassavica foliis salignis.—HERM., Parad. Bat. COM-
MELIN, Hort. Amst., p. 115, cum. ic.
Alnus maritima myrtifolia coriariorum.—PLUK., Almag., 18, t. 240, f. 8.
Alni fructu laurifolia arbor maritima.—SLOANE, Jam. Hist., ii. p. 18, t.
161, f. 2.
Innominata.—PLUM., ic. 135, t. 144, f. 2.

THIS is another tropical West Indian tree which the southern extremity of East Florida has afforded. It has been observed on the shore of Key West, Southern Florida, and around Tampa Bay. In the West Indies, like the Mangrove, with which it grows, and for a kind of which it is taken by the Spaniards, who call it *Mangle Saragoza*, it affects the low sandy and muddy shores near the sea, where it becomes an erect tree about thirty feet high, with the trunk a foot in diameter, having a smooth, whitish-gray bark and angular branchlets. In South America it also exists on the coast of Guayaquil, and in Chili, near Valparaiso. In a country where the finest kinds of wood are so common, that of the Button Tree is little esteemed, and it is, therefore, only used for fuel; it is, however, fine and close-grained, in the branches brownish white, capable of a high polish, with scarcely any visible annual layers, and made up almost wholly of dotted medullary rays. The general aspect of its inflorescence, and, indeed, its closely-imbricated inelegant heads of flowers, lead us almost to compare it with some of the *Amentaceæ*, particularly the Alder, while its real relations are to the present family, which includes in the *Combretum* itself, and the singularly-splendid *Cacoucia* of Aublet, some of the most elegant and beautiful of plants.

The bark is gray, bitterish, and astringent, and no doubt medicinal. The leaves, of a yellowish green, are from two to three inches long, three-quarters to an inch broad, acute at each end, very smooth, and on short petioles, which have frequently two glands at the base. The flowers, for which butterflies have a great predilection, are very inconspicuous, greenish yellow, small, and collected into globose heads, in axillary and terminal few-flowered panicles on pedicels about the length of the capituli. The heads at length become reddish; the capsules are small and scale-like, corky, dilated elliptic, internally concave, with broad, thin, carinated margins, and are very often abortive, never more than 1-seeded, and pubescent at the summit.

The island of Cuba affords another allied but very distinct species, which may be the *C. procumbens* of Jacquin, put down as a variety of the present by Decandolle; the calyx, however, is almost entirely smooth, with very acute segments, and the leaves are sharply apiculated, and sometimes obtuse with a short point. In this the wood appears to be very hard, and as close-grained as mahogany, of a dull white, inclining to gray, with a delicate feathered appearance, and a thick bark, gray externally and blackish within.

According to Prince Maximilian, the bark of the *Conocarpus racemosa* (one of the plants called Mangrove in Brazil) is much used at Rio Janeiro for tanning.

PLATE XXXIII.

A branch of the natural size. a. The flower. b. The fruit, of the natural size. c. The same magnified.

SILKY BUTTON TREE.

CONOCARPUS ERECTA. γ SERICEA, (Forster, in herb. l'Hérit.) *Foliis oblongis utrinque acuminatis ad utramque paginam etiam adultis adpressè villoso-sericeis.*—DECAND., Prod., vol. iii. p. 16.
Mangle foliis oblongis integris utrinque molli lanugine holosericea obductis.— SLOANE, Hist. Jam., vol. ii. p. 67, tab. 187, fig. 2.

ACCORDING to Sloane, this tree is known in Jamaica by the name of the White Mangrove, and attains the height of twenty feet, having white wood with a very small pith; the bark is also smooth and whitish. This variety, or species, has also been found, with the above, at Key West, in East Florida, by Dr. Blodgett. We do not see any thing to distinguish it as a separate species from the *erecta* except the peculiar silky, shining pubescence with which the leaves continue to be clothed even in the adult state.

LAGUNCULARIA.*

(GÆRTNER.)

Natural Order, COMBRETACEÆ. *Linnæan Classification*, DECANDRIA, MONOGYNIA.

Calyx border persisting, subcampanulate, 5-lobed. *Petals* five, minute, spreading, and caducous. *Stamens* five or ten, in two series, included. *Style* subulate; *stigma* capitate. *Nut* margined, coriaceous, valveless, 1 or 2-seeded, crowned with the calyx. *Cotyledons* convolute; the radicle very long.

A tree of the Caribbean Islands, with opposite, elliptic, smooth leaves. Racemes opposite, many-flowered, the flowers sessile, furnished with deciduous bractes; the calyx bibracteolate at the summit. The seed germinating within the nut. A genus nearly allied to LUMNITZERA of India.

WHITE MANGROVE.

LAGUNCULARIA RACEMOSA. *Gærtner*, fil. Carpol., vol. iii. p. 209, t. 217. DECAND., Prod., vol. iii. p. 17.
CONOCARPUS RACEMOSA. *Foliis lanceolato-ovatis obtusiusculis, fructibus segregatis.*—LINN., Sp. pl. WILLD., vol. ii. p. 995. SWARTZ, Obs., p. 79. JACQ., Amer., p. 80, t. 53.

---

* From *laguncula*, a little bottle, in allusion to the form of the nut.

Laguncularia racemosa

White Mangrove. Manglier à grappes.

WHITE MANGROVE.

Conocarpus, foliis elliptico-ovatis, petiolis biglandulosis, racemis laxis, fructibus sejunctis.—BROWNE, Jam., p. 159.
Mangle julifera, foliis ellipticis ex adverso nascentibus.—SLOANE, Jam., p. 156. Hist., vol. ii. p. 66, tab. 187, f. 1. RAII, Dendr., p. 115.
SPHENOCARPUS.—RICHARD, Anal. Fr., p. 92.
SCHOUSBOA *commutata.*—SPRENGEL, Syst. Veget., vol. ii. p. 332.

THIS plant is a native of the sandy and muddy shores of the Caribbean Islands and the neighboring continent, where it becomes a lofty, branching tree, sometimes dividing into three or four trunks close to the ground: it is called White Mangrove by the English inhabitants of these islands. Dr. Blodgett has sent specimens of this tree also from Key West, in East Florida.

The branches are cylindric and brownish, the twigs ferruginous; the leaves are opposite and smooth, about three inches long and an inch and a half wide, upon short petioles, quite entire, thick, and somewhat coriaceous, elliptic or ovate, obtuse, and sometimes emarginate, with a pair of glands near the summit of the petiole, and, in most of the leaves, toward the edge appear a number of raised glandular points, which are closed or open. The flowers are disposed in axillary and terminal elongated racemes, the racemes sometimes trifid. Flowers small, sessile, greenish white; the germ pyriform, and, as well as the short border of the calyx, covered with a short, whitish, silky pubescence. Petals five, very small and caducous. Stamens five, not exserted. The germ at its summit with two small, dentiform bracteoles, the bractes themselves short, broad-ovate, and caducous. Style, at length somewhat exserted, with a small, capitate stigma. Nut 1-seeded.

PLATE XXXIV.

A branch of the natural size. a. The nut in an early stage.

RABBIT BERRY.

Natural Order, ELÆAGNEÆ, (Jussieu.) *Linnæan Classification*, DIŒCIA, OCTANDRIA.

SHEPHERDIA,* (NUTTALL.) HIPPOPHAE (PURSH.)

Flowers DIŒCIOUS.—MALE calyx 4-cleft, much larger than that of the female. *Corolla*, none. *Stamina* eight, alternating with a torus of eight glands. FEMALE flower with a small 4-cleft, superior, campanulate *calyx*, and eight glands. *Style* one; *stigma* oblique, subcapitate. *Berry* juicy, 1-seeded, globose, invested with the fleshy calyx.

Small trees, spinescent or unarmed, with the general aspect of *Elæagnus*. Leaves entire, opposite, clothed with silvery and ferruginous scales. Flowers small, in axillary clusters, or in spikes. Berries pulpy, diaphanous, scarlet, subacid.

RABBIT BERRY,
OR
WESTERN SHEPHERDIA.

SHEPHERDIA ARGENTEA. *Foliis oblongo-ovatis, obtusis, glabris, utrinque argenteo-lepidotis, floribus glomeratis.*—NUTT., Gen. Amer., vol. ii. p. 240. LOUDON's Encyc. Plants, p. 836, Arboretum et Frutic., p. 1321, fig. 1208. HOOKER, Flor. Bor. Am., vol. ii. p. 138, tab. 178, (well illustrated.)
HIPPOPHAE *argentea.*—PURSH, Flor. Bor. Am., p. 115.

---

* Named in honor of the late Mr. Wm. Shepherd, then curator of the Liverpool Botanic Garden. A most scientific gardener and skilful cultivator.

Shepherdia argentea.

Rabbit Berry　　　　　　　　　　　Argousier argenté

This very useful, hardy, and ornamental tree is wholly a native of the northern and western regions of North America. Dr. Richardson observed it on the banks of the Saskatchawan, between Carlton and Edmonton House Forts, in the latitude of 54°, and Major Long's party found it growing on the borders of Rainy Lake, about latitude 49°. On the banks of the Missouri, the limit of its southern range is the borders of the Platte, but it appeared to be most abundant and fertile around Fort Mandan, or the Great Northern Bend of the Missouri, in about the latitude of 48°; here it becomes a small tree twelve to eighteen feet in height, and when adorned with its brilliant scarlet berries, produced in thick clusters so as almost to conceal the branches, few objects are more ornamental, contrasted also with the silvery hue of the leaves, which reminds one of the useful Olive: it presents at once an appearance both striking and novel.

Among the natives and Canadian voyageurs it has several different names. According to Lewis and Clarke, it was known on the Missouri, to the natives, by the name of the Rabbit Berry, from being fed on, probably, by those animals, and it was met with by their party in the gorges of the Rocky Mountains. Richardson says the Creek Indians give it a name which signifies Bloody Berry, (Methcoo-meeva,) from the singular redness and transparency of its fruit. The Canadian voyageurs called it *Graisse de bœuf*, or Buffalo-fat, from the imaginary relish of the berries, or from the practice of mixing them with their fat pounded meat or pemmican.

In 1815 I introduced a plant into the Liverpool Botanic Garden; but, being kept in the greenhouse, it was, I presume, killed with kindness, and was soon lost. About twelve or fifteen years ago, my friends Messrs. Windships, of Brighton, in Massachusetts, in the vicinity of Boston, obtained a few seeds of the *Shepherdia* from the banks of the Missouri, which, growing, gave origin to all the plants now in the United States and in Europe. A standard tree at this time growing in their nursery is quite as large as an ordinary Apple Tree of the same age, and it is yet very vigorous and increasing in size. It is, I believe, about twenty feet high, with a handsome, erect trunk, clothed with a somewhat smooth bark, and presents an appearance of the most vigorous health. Its resemblance to the Olive is very striking; it has a rounded, uniform, elegant summit, and when in fruit, which is at the close of summer, scarcely any thing can be more brilliant, from the load of berries with which it is everywhere clad; these are about the size of small red currants, juicy, but not watery, of a pleasant subacid taste, mixed with a sweetness which renders them generally agreeable. Made into sweet jelly, in the manner of currants, they are thought preferable by most who have tasted them. But the great use of the Shepherdia will be for constructing hedges or live fences, at least in the Northern States where it thrives well. Kept down by cutting, it becomes sufficiently close, and has also the advantage of being thorny, green, or rather silvery, till late in autumn, and it is attacked by no insect, nor subject to any disease or blight.

The berries are greedily devoured by all the autumnal birds, particularly robins and bluebirds, who flock round the tree in throngs while any thing remains to be had.

In its native state it is a small, rather narrow-topped tree, with the branches ending in stout spines. The leaves are oblong-ovate, obtuse, shortly petiolate, on both sides free from hairs, but covered with peltate or rounded scales, which (through a lens) appear to be ciliated. The flowers, which come out as early as in March, are in clusters. The calyx of the *male* flower is considerably larger than that of the female, and divided down to the base into four subovate, obtuse divisions, internally yellowish, but outside scaly like the leaves. The stamens are eight, with oftentimes rather short, pubescent filaments: the anthers are oblong and 2-celled.

The *female* flowers are smaller and shortly pedunculate, without any vestiges of stamens. There is one style, and a thickish, oblique, subelliptic stigma. The germ appears inferior, but is, in fact, only invested by the tube of the calyx. The berries are collected into clusters, and are sparingly scattered with scales, but bright and pellucid. The seed, or rather nut, with a cartilaginous shell, is subovate and shining, much like that of *Hippophae:* it is also scored externally, as if partly 2-lobed, with a small projection at the base. The embryo is straight and flat, without albumen, and the radicle inferior. The cotyledons are large, thick, and oval. Mr. Wyeth, in the Rocky Mountains, observed a variety of this species with yellow berries.

PLATE XXXV.

A branch of the natural size.

A second species of this genus is the

SHEPHERDIA CANADENSIS, or CANADIAN SHEPHERDIA, with elliptic-ovate leaves, nearly smooth above and naked beneath, clothed with stellate hairs and ferruginous scales: the flowers are also in axillary spikes. This species, it appears from Hooker, ranges far to the north, throughout Canada to Fort Franklin, on Mackenzie River, and from Newfoundland and Hudson's Bay to the Rocky Mountains. In the State of New York I have met with it on the borders of the smaller inland lakes, as well as on the shores of Lake Erie; but this species scarcely becomes a tree, is without thorns, and the berries, though equally brilliant with the preceding, are rather unpleasant to the taste. On the authority of Menzies, it is said also to exist on the northwest coast; but I have not seen it there.

XIMENIA.
(PLUMIER.)

Natural Order, OLACINEÆ.   *Linnæan Classification*, OCTANDRIA, MONOGYNIA.

Calyx 4-toothed, minute, persistent, not enlarging. *Petals* four, densely pilose within, connivent, above revolute. *Stamens* eight, the filaments capillary, anthers adnate, long, and linear, not exserted. *Ovary* 4-celled, 4-seeded. *Style* one. *Drupe* ovate, 1-seeded.—DECAND., vol. i. p. 533.

Tropical trees or shrubs, with smooth, alternate, entire, elliptical or ovate, exstipulate leaves; flowers mostly in small, axillary umbels.

MOUNTAIN PLUM.

XIMENIA AMERICANA. *Spinosa, foliis oblongis, pedunculis multifloris.*—LINN., Sp. pl. Hort. Cliff., 1193. Swartz, Obs., p. 149. DECAND., Prod., vol. i. p. 533.

α *ovata*, foliis ovatis. *X. multiflora.*—JACQUIN, Amer., p. 106, t. 277, fig. 31. LAMARCK, Illust., tab. 297, fig. 1.

XIMENIA *montana.*—MACFADYEN, Flora of Jamaica, p. 121; a variety, however, without thorns.

XIMENIA *aculeata, flore villoso, fructu luteo.*—PLUMIER, Gener., p. 6. Ic. 261, fig. 1.

* Named in honor of Francis Ximenes, a Spanish naturalist and missionary.

Ximenia Americana.

Mountain Plum Ximenie Americaine

MOUNTAIN PLUM.

This plant forms a small tree, with an erect stem and spreading, gray, verrucose, and somewhat angular branches. It is indigenous to the mountains of Jamaica, Key West in Florida, where it was found by Dr. Blodgett, and is also met with in the neighborhood of Carthagena, in Hispaniola, and many years ago it was collected in the interior of East Florida by John Bartram, as Mr. A. Gray saw specimens of it in his collection still extant. According to Drs. Wight, and Roxburgh, it is also indigenous to the coast of India.

It bears a drupe the size of the plum of Europe, or of a pigeon's egg, yellow, smooth, and shining, 1-seeded, with a thin rind and watery pulp of a pleasant sweet subacid taste. The seed is large and white. This plum is of an agreeable flavor, and not inferior to the common varieties of that of Europe; it has a slight degree of astringency, with a pleasant acidity. The flowers have a fragrant odor said to be like that of frankincense. The wood is as yellow as that of the Sandal, and, in India, its powder is often substituted for it by the Brahmins in their religious ceremonies.

The leaves grow two or three together, on short, lateral, tuberculoid branchlets; they are petiolate, oblong-lanceolate, obtuse, and narrowed below, smooth, obscurely veined, about two or two and a half inches long, and less than an inch broad. The flowers are disposed in small pedunculated axillary and subterminal umbels, the umbels three or 4-flowered. The calyx is minute and 4-toothed. Petals four, linear-oblong, conniving into a tube below, recurved at the apex, and covered with rather long and dense brownish-yellow hairs within. Stamens eight, as long as the petals, the filaments like the most delicate threads, the anthers long and linear, ovary 8-angled at the base, conical and subulate, with the style as long as the stamens.

PLATE XXXVI.

A branch of the natural size. a. The fruit.

MACLURA.*

(NUTTALL.)

Natural Order, ARTOCARPEÆ, (R. Brown.) *Linnæan Classification*, DIŒCIA, TETRANDRIA.

Flowers DIŒCIOUS.—MALE flowers with the *calyx* 4-cleft. *Stamens* four, exserted. FEMALE flowers in globular aments; the *calyx* fleshy, 4-parted, with the bases ingrafted together. *Corolla* none. *Style* one, filiform, villous. *Germs* numerous, each 1-seeded, coalescing into a compound globular berry of many cells, the cells 1-seeded. *Seed* obovate.

A lactescent tree, related to the Fustic and with nearly similar yellow wood. Leaves alternate, entire, without stipules, producing superaxillary simple spines. Male flowers in pedunculated umbels. Aments axillary. Berry verrucose and large, resembling an orange, at first lactescent, greenish-yellow or yellow. Nearly allied to *Broussonetia*, but distinguished by the coalescence of the germs, and a peculiar habit.

OSAGE ORANGE, OR YELLOW WOOD.

MACLURA AURANTIACA.—NUTT., Gen. Amer., vol. ii. p. 234. LOUDON'S Encycl. of Plants, p. 784, fig. 13256. Arboretum et Frutic, vol. iii. pp. 1342, 1362, figs. 1226, 1227, 1228. LAMBERT'S Supplement to Pinus, l. c. EATON'S Manual, (Ed. 8,) p. 311.

* In honor of the celebrated geologist, philosopher, and patron of natural science, Wm. Maclure.

Maclura Aurantiaca.
Osage Orange. Bois d'Arc.

OSAGE ORANGE, OR YELLOW WOOD.

This remarkable tree, though perfectly thriving and hardy in the climate of Philadelphia, raised from seeds, does not naturally extend to the north of the Arkansas River; it is even there of rare occurrence, and generally destitute of fruit. We saw a few old ill-grown trees on the banks of the Pottoe, a few miles from Fort Smith. It was only on the rich low bottom-lands of Red River, near the confluence of the Kiamesha, that we beheld the Maclura in perfection, forming a great part of the prevailing umbrageous forest, and attaining an elevation of fifty or sixty feet by a diameter of between two and three feet. It is, however, much inclined, when full-grown, to throw out irregular spreading limbs of considerable extent, though at first it presents a very elegant roundish spreading summit. But at all times it strikes the beholder as something remarkable in the Northern forest by the beauty and splendor of its dark and shining foliage, which in appearance strongly resembles that of the Orange; and the numerous spines which the branches present seem to confirm the comparison. The fruit is alike singular, resembling full-grown yellowish-green oranges, but filled with a milky juice, and, as they ripen at intervals, or become abortive, the ground beneath, like an orchard, is strewn with this curious fruit, which, when first discovered lying neglected beneath the tree, led the voyagers to fear and report it as a poison; but the family to which it appertains, and its relation to the Mulberry, afford a presumptive sanction for its harmless qualities.

The Maclura was first noticed by Hunter and Dunbar in their voyage up Red River, on the banks of the Little Missouri, of the Washita River; also near Natchitoches, and upon the banks of the Arkansas. It was likewise observed by Dr. James, in Major Long's expedition, along the banks of the Arkansas and the Canadian. I first saw living plants, bearing fruit, about the year 1810, in the garden of Mr. Chouteau, at St. Louis, which were raised from seeds collected in the country of the Osages. It was afterward introduced into the garden of the late Mr.

McMahon, of Philadelphia, from seeds collected by Governor Lewis. The largest tree I have seen in cultivation is in the garden of Mr. Thomas (now David) Landreth, in Federal Street, Philadelphia. It is about twenty feet high and five and a half in circumference. This tree has for many years borne fruit; but the seeds have only of late been perfected, in consequence of the absence of the staminiferous individual, on the accession of which, however, it produced abundance of seeds nearly as large as those of the Orange, and much of the same form: these were indicated in the fruit by an unusual projection of the external tubercles.

The wood of the Maclura is solid, heavy, and elastic, of a fine yellow color, which, like the Fustic, it readily communicates to water, and it might consequently be used as a dye. It is also capable of receiving a fine polish, and resembles satin-wood by its brilliant gloss. The elasticity and durability of its wood have long been celebrated by the Indians, who, from its use, have bestowed upon it the name of Bow-wood, and the Canadians, who traversed these regions in quest of furs, knew it long by the name of the *Bois d'arc*. The bark, as in *Broussonetia*, affords a fine white flax. Another important use of the Maclura, in this climate, is that of forming live fences or hedges, for which purpose it is well adapted, as it bears cutting, grows close, and is very thorny, as well as free from the attacks of blight and insects: it has all the advantage of keeping for a long time green, and appears, in all respects, as elegant a fence as that of the Wild Orange in the South. It has besides an additional recommendation in its use for feeding silk-worms, for which purpose it is scarcely inferior to the famous *Morus multicaulis*.[*]

---

[*] Different opinions are now entertained of the value of the leaves of the Maclura as a food for silk-worms, some approving and others discouraging their use.

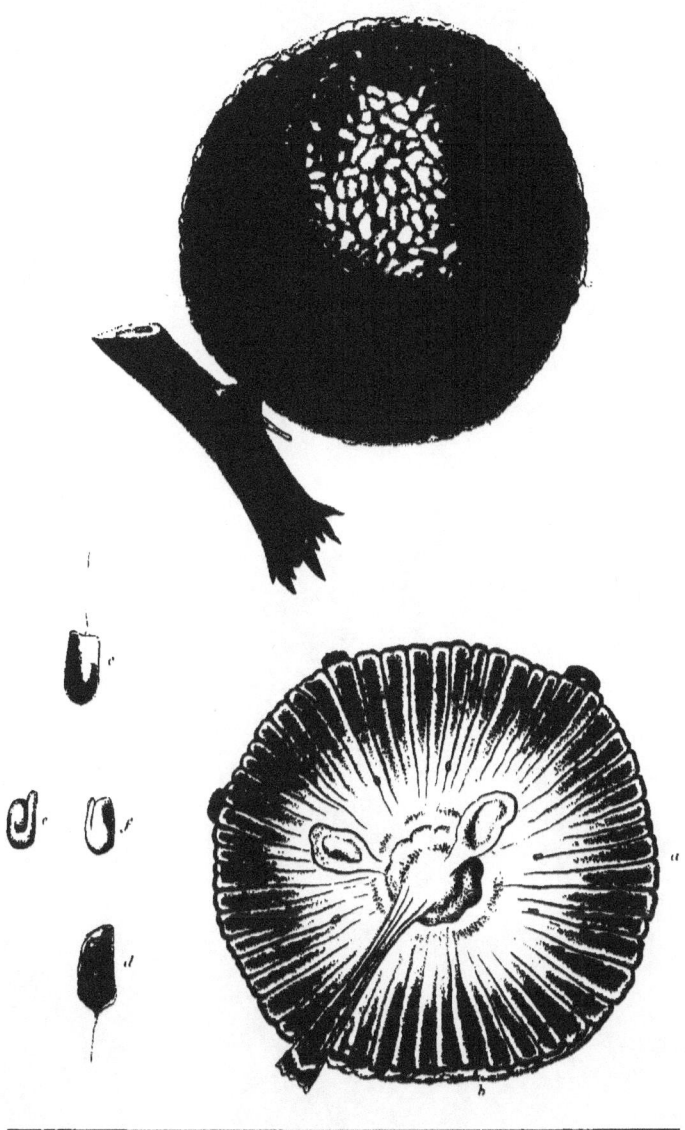

Maclura Aurantiaca.

Osage Orange. Bois d'Arc

OSAGE ORANGE, OR YELLOW WOOD. 143

The branches are flexuous and round, clothed with a smooth gray bark. The leaves are alternate, upon long foot-stalks, and are usually oval and acuminated; on the bearing branches they are, however, often considerably larger, and heart-shaped at the base, very entire, with the point mucronated and a little pungent; the upper surface is smooth and shining, but the petiole and nerves on the under side of the leaf are somewhat hirsutely pubescent. The petiole is often an inch or more long; the leaf itself two to four inches, and one and a half to three inches wide. The *staminiferous* plant appears uniformly weaker, more delicate, and smaller-leaved, than the fertile plant. The flowers in it are axillary, in pedunculated small umbels, each umbel containing about fifteen to twenty flowers, consisting merely of a small 4-cleft calyx, with oval hairy segments, and four stamens, on lengthened and exserted filaments. The anthers are 2-celled, large and oval, opening lengthwise. In both plants, single, undivided thorns come out in the upper axils of the leaves. The *female* capitulum consists of a congeries of flowers united into a globular form, about the size of a cherry; these consist also in a calyx of four divisions, but less regular than in the male. The styles and stigmas, one to each germ, are three-fourths of an inch long, giving to the ament the appearance of a tuft of long pubescent threads. The berry, filled with a milky juice, becomes about the size of a moderate but not large orange, having an irregular tessellated appearance, almost like that of the Bread-fruit: these tessellations are the unduly-enlarged fleshy summits of the segments of the calyx. The whole of the calyces, at a short distance below the immediate surface, become partly ingrafted together into one mass; and a transverse section of the fruit, therefore, presents a series of radiating and woody fibres, among which are immersed the 1-seeded germs. The seeds are obovate, compressed, and nearly as large as those of the orange. The testa is yellowish white and cartilaginous; there is no albumen, and the seed is of a yellowish brown, pendulous, with the radicle inverted from the axis of the fruit, and curved partly over the margin of the thick and fleshy cotyledons toward the point occupied by the hylum. The fruit, when ripe, is succulent, has a sweetish but insipid taste, and is somewhat acrid. As far as we know, it is not eaten by any animal.

It is readily propagated by sowing the seeds, and also by cuttings of the root, which grow much more readily than cuttings of the branches. Although several male plants were raised in this vicinity, it is singular to remark, that while the fertile plant is in its utmost vigor, very few of the former survive; and, as their presence is necessary for fertility, it probably would be the best way to ingraft the male on some one of the branches of the bearing plant.

PLATE XXXVII.

A branch of the male plant of the natural size. a. The male flower a little enlarged. b. The female capitulum.

PLATE XXXVIII.

The fruit of the natural size. a. A transverse section of the same. b. The appearance of the abortive germs in the section. c. The seed with the testa. d. The same divested of the testa. e. The embryo. f. The back view of the same.

Besides the Fig, we have, in this interesting family of plants, the famous Bread-fruit of India and the islands of the Pacific; the Cow Tree, or Palo de Vaca of South America, which is tapped like a fountain, and yields a copious supply of rich and wholesome milk. The nuts of the *Brosimum alcicastrum*, when roasted, are used in the place of bread, in the West Indies, and have a taste similar to that of hazel-nuts: the juice of all the plants of this family also contains more or less of caoutchouc. Amidst this generally harmless group of plants, it is singular to find the deadly Upas of Java, (*Antiaris toxicaria,*) whose inspissated juice is found to contain that most virulent of all poisonous principles, strychnia. It is, however, somewhat doubtful what the real affinities of this plant are, as it is acknowledged to be an anomaly in the family.

NETTLE TREE.

Natural Order, ULMACEÆ, (Mirbel.)   *Linnæan Classification*,
POLYGAMIA, DIŒCIA.

CELTIS.* (LINN.)

Flowers POLYGAMOUS.—*Staminate* flower with the calyx 5 or 6-parted. *Corolla* none. *Stamens* four to seven. *Perfect flower* with the calyx deeply 5-parted. No *corolla*. *Ovarium* 1-celled; the *ovules* solitary and pendulous. *Style* very short; *stigmas* two, thickish, subulate, and spreading. *Drupe* globular, 1-seeded, thinly coated with a sweetish pulp. *Embryo* inverted. *Cotyledons* folded.

The genus within its proper bounds includes deciduous-leaved trees of South Europe, the Levant, the mountains of Nepaul and Cuba, and the forests of the United States. The true Celtides have alternate, entire, deciduous, and mostly cordate leaves, generally oblique at the base, 3-nerved, entire, but mostly serrated on the margin. The stipules are membranaceous and deciduous. The flowers are precocious, or appear before the expansion of the leaves, with a film-like irregularly-torn membranous perianth, the staminiferous ones near the base of the branchlet pedicellate, and three or four together. The fertile flowers are solitary and axillary, on short peduncles. The drupes brownish yellow, rather sweet, insipid, and nearly juiceless.

For the tropical species with axillary cymes coeval with the leaves, two distinct styles, and an ovary with two ovules, I propose the name of TRACHYDENDRON, (in allusion to their rough pubescent leaves and twigs.) Most of these species have a tough fibrous bark of the nature of hemp.

---

* The ancient name of the *Lotus*, applied to this genus by Linnæus.

Celtis reticulata.

Small leaved Nettle tree. Micocoulier réticulé

SMALL-LEAVED NETTLE TREE.

CELTIS RETICULATA. *Foliis brevibus, lato-cordatis, subcoriaceis, vix et irregulariter serratis acutis basi obliquis scabris, subtus subglabris venis elevatis reticulatis, pedunculis fructiferis unifloris.*

CELTIS RETICULATA.—TORREY, in Annals of Lyceum, N. Y., vol. ii. p. 247.

THIS low-growing species of Nettle Tree was discovered by Dr. James near the base of the Rocky Mountains; I likewise met with it in the same mountain range, by small streams, and also along the borders of the Oregon, toward the Blue Mountains, particularly along the banks of the Brulée, a small stream falling into that river. It does not, in the situations where we observed it, become a timber tree, but rather a tall shrub, full of slender, and at length smooth, branches. The leaves become thick and rigid, and are about an inch and a half long by less than an inch wide, acute, but scarcely acuminate, with a few irregular serratures toward the point of the leaf, though a number of the leaves may be observed possessing no serratures at all; the upper surface is shining and scabrous; beneath the leaves are pubescent along the nerves, though at length nearly quite smooth; the petioles are one or two lines long and pubescent; the base of the leaf is very oblique, rounded, and slightly sinuated. The drupe is globose, solitary, brownish yellow, on a short peduncle. Of the wood of this species nothing is yet known.

PLATE XXXIX.

A branch of the natural size.

LONG-LEAVED NETTLE TREE.

CELTIS LONGIFOLIA. *Foliis ovato-lanceolatis promisse acuminatis integerrimis lævigatis demum glabris basi rotundatis obliquis subcuncatis, pedunculis fructiferis unifloris, cortice sublævi.*
CELTIS *Occidentalis, β integrifolia.*—NUTT., Gen. Am., vol. i. p. 202, (not of LAMARCK.)

THIS tree, growing to the height of sixty or seventy feet, inhabits the deep shady forests which border the Mississippi from St. Louis to the vicinity of the sea. Its even and not deeply cleft bark, in the absence of its aspiring summit, at once distinguishes this species from the Common Hack Berry. Like all the rest of the genus, (confined within its proper limits,) the insignificant filmy flowers appear early in the spring, before the expansion of the leaves. The small branches are smooth and yellowish brown. The leaves are smooth, of a thin consistence, and remarkable for the great length of their acuminated points, of an ovate or ovate-lanceolate form, subtended by deciduous stipules, and at first pubescent beneath, particularly along the large vessels or veins; at length almost absolutely smooth, at no time scabrous either above or beneath, and wholly entire on the margin. The length is about from three to three and a half inches by one to one and a half wide. The obliquity of the base varies according to the position of the leaf on the branch: those first developed are nearly equal at base, and of a lanceolate form; the later ones are larger, wider, and more oblique. The flowers are as usual; the males about three together toward the base of the branch, the females solitary and axillary, coming out with the opening leaves. The stamens are from five to seven. The margin of the calyx-segments is pubescent, but smooth, narrower, and more deeply divided in *C. Occidentalis*, to which this species

Celtis longifolia.

Long leaved Nettle tree . Micocoulier à longues feuilles.

bears a close affinity. The berry is of a brownish yellow, on a short peduncle.

PLATE XL.

A branch of the natural size.

THIN-LEAVED NETTLE TREE.

CELTIS TENUIFOLIA. *Foliis lato-ovatis acuminatis subæqualiter serratis basi inæqualibus subcordatis utrinque glabriusculis junioribus pubescentibus, calycibus laciniis cucullatis crosis ciliatis, fructibus solitariis.*—NUTT., Gen. Am., i. p. 202.

CELTIS *Occidentalis*, β? *tenuifolia.*—LAM., Encyc. Bot., pp. 137, 138. PERSOON, Synops., i. p. 292.

CELTIS *Occidentalis.*—DARLINGTON, Flora Cestrica, p. 180.

CELTIS *pumila.*—PURSH, Flor. Bor. Am., i. p. 200 and 201.

THIS species, nearly allied to *C. Occidentalis*, is often rather a shrub than a tree. Near Harper's Ferry, in Virginia, on the summits of the neighboring mountains of the Blue Ridge, it flowers at the height of two feet, with low spreading branches. In other situations it becomes a tree, as in the vicinity of Bethlehem, in Pennsylvania, and other places in that State. It is best distinguished by its broad, rather short, and smooth leaves when adult, and which are not scabrous on the upper surface: the margin is now and then without serratures; their form is cordate-ovate, very little acuminated. The berries are solitary, brown, and glaucous. It is, perhaps, little more than a variety of *C. Occidentalis*.

The AMERICAN NETTLE TREE (*"Celtis Occidentalis"*) occurs in almost every part of Massachusetts, and I know a tree of it, east of the Connecticut River, nearly five feet in diameter."—G. B. EMERSON, Esq.

THICK-LEAVED NETTLE TREE or HACK BERRY, (*Celtis crassifolia*.) This species is readily distinguished by the large size, thickness, and roughness of the leaves; they are often full six inches long by more than three inches wide. The flowers, which are very numerous, (in the trees of the forests of Kentucky,) appear long before the development of the leaves, and are subtended by broad caducous bractes. The divisions of the calyx are spathulate, cucullate, scarcely torn, and not ciliate. The fruit, as well as the flowers, from the falling of the small leaves which accompany it, often appears disposed in racemes. The drupes are dark brown, nearly black. This species also admits of some variety. In some the leaves are larger and more pubescent or even hirsute, while others are nearly smooth when adult.

The wood of the Nettle Tree of Europe is blackish, hard, and compact, heavy and without sap-wood. It is so pliable and tenacious, that it will bend much without breaking; it therefore makes a good wood for the shafts of carriages and other uses of the wheelwright. It forms hoops which are very durable; and it is said that, after ebony and box, it surpasses all others in durability, strength, and beauty. It is likewise unattacked by worms; and is esteemed for works of sculpture, as it neither contracts nor cracks. The roots serve as a dye for linen stuffs; and an oil has been obtained from the seeds very similar to that of sweet almonds.

Ficus pedunculata.

Cherry fig-tree. Figuier pédunculé.

FIG TREES.

Natural Order, ARTOCARPEÆ, (R. Brown.) *Linnæan Classification,* POLYGAMIA, DIŒCIA.

FICUS.* (TOURN. LINN.)

DIŒCIOUS.—The common receptacle spherical or pyriform, resembling a berry, fleshy and closed, including numerous distinct and minute flowers. *Male,* calyx 3-parted. *Corolla* none. *Stamens* one to three. *Female* with the calyx 3 to 5-parted, and no corolla. *Pistillum* one; style one, subulate; the stigma simple or bifid and unequal. *Seed* one, covered by the persistent subcarnose calyx.

Lactescent trees or shrubs, chiefly of Tropical America, Africa, and India; leaves alternate, stipulate, stipules terminal, conical, convolute. Receptacles mostly axillary, solitary, or crowded, rarely disposed in terminal racemes, often bracteolate at base.

CHERRY FIG TREE.

FICUS PEDUNCULATA. *Foliis ovato-oblongis integerrimis acuminatis obtusis, basi obsolete cordatis, receptaculis globosis subgeminatis calyculatis pedunculatis.*—WILLD., Sp. pl., AITON, Hort. Kewen., vol. iii. p. 450.
Ficus arbor Americana, arbuti foliis non serrata, fructu pisi magnitudine,

---

* A Latin word of uncertain derivation.

funiculis e ramis ad terram dimissis prolifera.—PLUKEN., Almag., p. 144, tab. 178, fig. 4.?

THIS species of Fig Tree was discovered by Jacquin in the island of Martinique; it is also indigenous to some other of the West India Islands, as well as to the neighboring continent of Tropical America. At Key West, according to Dr. Blodgett, it becomes a large spreading tree fifty feet in height, and, like some other species, particularly the famous Banyan Tree, (*F. Indica*,) it sends down roots from its lofty branches resembling ropes, which, on reaching the soil, at length become so many independent trunks, in turn producing others; and, spreading themselves on all sides without interruption, they present a united summit of prodigious extent, which, reposing on a multitude of trunks of different dimensions, seems like the airy vault of some vast edifice sustained by innumerable columns.

The bark of the branches appears to be gray and even; the leaves are very smooth on both sides, but covered with innumerable minute dots on the upper surface. They are three to four inches long, one and a half to two inches wide, with a peduncle about one and a half inches long. They have a few distant pennated nerves inosculating toward the margin of the leaf, with innumerable intermediate slender reticulations of vessels; they are generally of an ovate form, rounded or almost cordate at the base, with a short and blunt acumination; from their axils arise one or two peduncles about three-quarters of an inch long, each terminated by a bifid involucel, improperly called a calyx. The figs themselves are nearly globose, but sensibly wider at the summit, about the magnitude of small cherries, greenish-yellow and purple at the summit, (as they appear in a withered state,) with a few purplish pale spots.

Of this species there appears to be a distinct variety, if not a species, which I shall for the present call β *acuta;* the leaf is

Ficus brevifolia.

Short leaved Fig-tree.   Figuier à feuilles courtes.

elliptic, shortly acuminate, acute at base, and faintly nerved beneath. It also becomes a large tree, producing a fig about the size of a cherry, which is yellow when ripe.

PLATE XLI.

A branch of the natural size. a. The fruit.

SHORT-LEAVED FIG TREE.

FICUS BREVIFOLIA. *Foliis cordato-ovatis integerrimis, obtusis abbreviatis brevi petiolatis glabris, venis immersis, receptaculis globosis depressis umbilicatis solitariis brevi pedunculatis, involucellis bifidis.*

THIS is also a species of arborescent Fig, indigenous to Key West, in East Florida, but by no means common, and, according to Dr. Blodgett, its discoverer, it forms a tree with a slender, almost horizontal, stem, which in its progress throws off funicular roots, that serve as props to prevent the main trunk from becoming entirely prostrate.

The branches are covered with a light gray bark. The leaves are about two inches long by one and a half inches wide, perfectly smooth on both surfaces, on petioles from one-half to three-fourths of an inch long. The veins on the under surface are so far immersed as to be scarcely visible. The figs, about the size of small cherries, are of a flattened, spheroidal form, at first, as well as the bifid involucrum, slenderly villous; they grow out chiefly at the extremities of the twigs, on thick pedicels, about two or three lines long, and when ripe are of a brightish purple red. We do not find any species with which the present agrees. From the description and specific name,

IV.—10*

we should suppose the present species allied to the *F. padifolia* of Humboldt and Bonpland, but it differs too much to be referred to that species.

PLATE XLII.

A branch of the natural size.

SMALL-FRUITED FIG TREE.

FICUS AUREA. *Glabra, foliis integerrimis ellipticis subacuminatis acutiusculis basi plerisque angustatis pennincrviis brevi-petiolatis, fructibus globosis geminatis sessilibus involucratis, involucris subtrifidis majusculis.*
β LATIFOLIA. *Foliis lato-ovatis subellipticis.*

THIS species, according to its discoverer, Dr. Blodgett, becomes, at Key West, in East Florida, a large tree, at first parasitical on other trees, but, destroying its supporter, it at length reaches the ground and forms an independent trunk of large dimensions. It bears, however, a very insignificant fruit, only about the size of a pea, and orange-yellow when ripe.

The branches are covered with a whitish bark. The leaves, three to four and a half inches long, are from one and a half to two and a half broad; the peduncles are about an inch in length. The form of the leaves is almost similar with those of the Orange, elliptic and narrower below, with a rounded summit, and a very short, rather acute, projecting point or acumination; they are dark green above, paler beneath, with conspicuous feathered nerves which run together within the margin. The figs are sessile, clustered by pairs, with a small, valvular orifice, and are nearly half embraced by the sheathing, bifid or trifid, one-sided involucrum.

Ficus Aurea.

Small Fruited Fig-tree. Figuier doré.

BANYAN TREE.

This species appears to be very nearly allied to the *F. Martinicensis* of Willdenow, (the *F. laurifolia* of Lamarck;) but we can by no means reconcile it to Sloane's figure, (Hist. Jam., t. 223,) for in that species the leaves are lanceolate, and eight or nine inches long by two wide, on petioles two inches or more in length; the fruit is also said to be scarlet, of the size of a hazel-nut, and sweetish and not unpleasant. In our variety β the leaves are wholly oval and not narrowed at the base.

PLATE XLIII.

A branch of the natural size.

The milky juice of the Fig Tree is more or less acrid and fetid, however sweet and wholesome the fruit may be, and that of the *Ficus toxicaria* of Sumatra is accounted poisonous. The sap of several of the South American and Mexican species, inspissated, affords caoutchouc.

The cultivated Fig, (*Ficus carica*,) in its wild state, is an humble and distorted shrub, affecting rocks and ruins, bearing a fruit of inferior flavor, but with the parts of fructification very perfect. Such figs as seem to drop off before maturity are commonly those in which the stamens are most numerous or effective. These are carefully collected in the Levant to fertilize the female blossoms of the cultivated Fig, which will explain the mystery of *caprification*. In these countries the fruit, fresh, or dried in the sun, forms an important part of the food of the inhabitants.

The BANYAN TREE, (*Ficus Indica*,) nearly allied to our *F. aurea*, becomes in India an immense tree, spreading very wide, and throwing down rope-like roots into the soil. Marsden mentions one of these growing near Memgee, twenty miles west of Patna, in Bengal, which gave the enormous diameter of 370 feet; the circumference of the shadow, at noon, was 1116 feet, and there were fifty or sixty stems. It is called the Priests' Tree, and is held in such veneration by the Gentoos, that if any one cuts off a branch, he is looked upon as sacrilegious, and held in the greatest abhorrence.

Cratægus sanguineus.
Red Thorn. Alizier rouge.

HAWTHORN.

Natural Order, ROSACEÆ, (*suborder*, POMEÆ, Jussieu.) *Linnæan Classification*, ICOSANDRIA, DI— PENTAGYNIA.

CRATÆGUS. (LINN., excluding some species.)
Adnate *calyx*-tube urceolate, with a 5-cleft border. *Petals* five, orbicular. *Stamens* fifteen or more. *Styles* three to five, (rarely one.) *Pome* or apple fleshy, ovate or globose, closed and crowned with the persisting teeth of the calyx, and containing two to five hard 1-seeded nuts.

These are spiny shrubs or small trees, almost exclusively indigenous to Europe and the United States of America, with simple, angularly-lobed, incised, or toothed leaves, furnished with stipules of somewhat different forms on the fertile or infertile branches. Flowers white, in terminal corymbs, sometimes solitary. Bractes subulate, deciduous. The fruit rather small, sweet, or agreeably acidulous.

RED THORN, OR SIBERIAN HAWTHORN.

CRATÆGUS SANGUINEA. *Spinosa, foliis septangulis serratis basi productis petiolis submarginatis.*—PALLAS, Flora Rossica, vol. i. p. 25, tab. 11, (very good.)
CRATÆGUS SANGUINEA. Leaves broadly obovate, somewhat cuneate at the base, incised and serrate, often slightly 5 to 7-lobed, a little pubescent when young, on short petioles, at length coriaceous and shining; corymbs glabrous or somewhat pubescent; segments of

the calyx entire, and, as well as the pedicels, not glandular; styles 3–4; fruit globose.—TORREY and GRAY, Flora N. Am., vol. i. p. 464.

β *Douglasii.* Spines short and stout, (long in cultivation, *Loudon;*) fruit small, dark purple.

CRATÆGUS *punctata*, β *brevispina.*—DOUGL., in HOOK. Flor. Bor. Am., vol i. p. 201.

CRÆTAGUS *glandulosa.*—PURSH, vol. i. p. 337, (as it regards the plant collected by Capt. Lewis in the Rocky Mountains.)

CRÆTAGUS *Douglasii.*—LINDL., Bot. Regist., tab. 1810. LOUDON, Arbor. Brit., vol. iii. p. 823.

THIS species of Hawthorn, which becomes a tree eighteen to twenty-five feet in height, is first met with to the West, on the borders of rivulets, in the range of the Rocky Mountains, particularly on their western declivity, from whence it continues along the banks of the Oregon, and particularly its tributaries, down to the shores of the Pacific. We found it also, in great perfection, loaded with its sweet, nearly black, and pleasant fruit, on the banks of the Wahlamet. The stem attains to about the diameter of three to six inches, with a whitish, compact, close-grained wood, of which, in common with the Crab Apple of that country, the natives make their wedges for splitting trees.

The Siberian plant, according to Pallas, begins to be found to the south of the Uralian Mountains, and continues beyond the Obi through all the southern tract of Siberia, in dry mountainous situations, and in the thickets which border the higher rivulets; exactly the sort of situations affected by the American plant in the alpine region where it commences. It also, like ours, becomes a tree twelve to eighteen feet in height.

Lewis and Clarke speak of finding haws, probably of this or the following species, on Flat Head River, which heads against the sources of the Missouri.

Almost entirely deprived of vegetable food, every accession of fruit, however meagre, was hailed with delight by our famished party, and the ripe berries of this fine Hawthorn were collected with avidity. The bushes, or rather trees, were, however, so high that we could only come at the fruit on horseback, or after ascending the trunk, which often appeared equal to that of an ordinary Apple Tree.

The summit of the tree is round and spreading, and the thorns vary in size, though they are often short, and in no case numerous. The leaves are broad and somewhat rounded above, cuneate at the base, smooth on the upper surface, and always more or less pubescent beneath; the margin is incise and serrate, and divided often into five to seven shallow lobes. The flowers are white, rather large, and numerous, disposed in a corymb, with the peduncles and base of the calyx more or less pilose and glandular. The styles, three or four, are occasionally as many as five. The segments of the calyx are rather long and acuminated, membranaceous on the margin, and appressed to the flower. The berries are shortly elliptic or oval, and nearly black or dark purple when ripe. In the Siberian plant, described by Pallas, they are scarlet; but he remarks that, according to Steller, the haws of Kamtschatka are both red and black, and that there they are not only used as agreeable fruits, but are also collected for the purpose of distillation into spirits. A good spirit is likewise obtained by the fermentation and distillation of the fruit of the common Hawthorn, (*C. oxyacantha.*)

This species is very nearly allied to *C. coccinea*, with which indeed Pallas compares it; but in *C. coccinea* the leaves have longer petioles, it bears much larger flowers, with larger segments to the calyx. The fruit is also (in our plant) smaller, and the plant more decidedly arborescent.

PLATE XLIV.

A branch of the natural size. a. *The fruit.*

RIVER HAWTHORN.

CRATÆGUS RIVULARIS. *Foliis ovatis vel obovatis, obtusis acutisve inciso-serratis basi attenuatis brevi-petiolatis; corymbis multifloris glabris, floribus parvulis, calycis laciniis obtusis brevissimis eglandulosis; fructibus nigris.*—NUTT., in TORREY and GRAY, Flor. N. Am., vol. i. p. 364.
β CUNEATA. *Spinis brevibus, foliis cuneatis obtusis, incisis.*

ALONG the shady borders of the rivulets of the Rocky Mountains we observed this species blended with the former, becoming equally a tree and producing the same kind of pleasant dark fruit. It was also observed by Douglas in the interior of Oregon, where we likewise met with it. It is, in all probability, the smoother, supposed variety of *C. punctata*, mentioned by Hooker in his Flora.

The branches are reddish brown, the leaves nearly as entire as those of the Apple Tree, except in β, where they are slightly lobed; beneath very smooth, slightly pubescent above, acute and rather sharply serrate, with long spines. The peduncles and calyx perfectly smooth, the segments of the latter mere broad, obtuse dentures. The flowers are white and smaller than in the preceding. The berries are also black, and possess nearly the same sweet and rather insipid taste of the Common Haw, (*C. oxyacantha*.)

LANCE-LEAVED HAWTHORN.

CRATÆGUS ARBORESCENS. *Inermis, foliis lanceolatis inciso-serratis utrinque acutis rariter sublobatis glabris subtus ad venis puberulis, corymbis multifloris, colicibus pilosis laciniis subulatis integris, floribus pentagynis.*
CRATÆGUS *arborescens.*—ELLIOTT, Sketch., vol. i. p. 550. TORR. and GRAY, Flor. N. Am., vol. i. p. 466.

Cratægus Arborescens.

Lance leaved hawthorn. Alizier arborescent.

LANCE-LEAVED HAWTHORN.

ACCORDING to Elliott, this species becomes a tree of twenty to thirty feet in height, with spreading branches. The fruit is globose, quite small, and red. Of the quality of the wood nothing is yet known; but nearly all the arborescent species are of slow growth, and have whitish, close-grained, very hard, and durable wood; that of the Common Hawthorn (*C. oxyacantha*) is tough, and in England is used occasionally for axle-trees and handles of tools.

The Lance-leaved Thorn of Mr. Elliott was found on the borders of the Ogeechee River, in Georgia, near Fort Argyle, and near New Orleans, and in Texas by Drummond and Berlandier. It is without armature. The leaves are lanceolate, acute at each end, deeply serrated, smooth both above and beneath, except some small tufts of hairs at the divisions of the veins, sometimes slightly lobed toward the summit, (though not at all in our specimen.) The flowers are small, the calyx hairy at the base, with the segments small and subulate.

To show the great age to which the Common Hawthorn attains, Withering states of the variety called the Glastonbury Thorn, existing in his time, in a lane by the churchyard of the abbey, (1801,) "It appears to be a very old tree. An old woman of ninety never remembers it otherwise than as it now appears. It blossoms twice a year: the winter blossoms, which are almost the size of a sixpence, appear about Christmas, and sooner, if the winter be severe. These produce no fruit." The summer flowers bore berries containing only a single seed, which, when sown, produced plants nowise differing from the common kind.

The Common Hawthorn, though so humble in the hedgerow beneath the cropping of the shears, when suffered to grow up and stand alone attains the ordinary size of an Apple Tree; and, occupying the village green for a long series of years, it becomes connected with our earliest recollections of the joyful arrival of spring. The *old Hawthorn*, again white with its fragrant blossoms, and their falling on the ground like a shower of snow, marks a delightful era in the distant reminiscences of the writer, when yet the simplest boon of nature gave delight. With these pleasing recollections of the past, how touching and graphic are those beautiful lines of Goldsmith descriptive of the "Deserted Village:"—

>"The Hawthorn bush, with seats beneath the shade,
>For talking age and youthful converse made!
>How often have I bless'd the coming day,
>When toil remitting lent its turn to play,
>And all the village train, from labor free,
>Led up their sports beneath the spreading tree."

PLATE XLV.

A branch of the natural size. · a. The germ and styles.

NAKED-FLOWERED HAWTHORN,

OR

APPLE HAW.

CRATÆGUS ÆSTIVALIS. *Subspinosa, floribus præcocibus, foliis oblongo-cuneiformibus vel ellipticis brevi-petiolatis apice subsinuato-dentatis angulatis vel inciso-crenatis rarius trilobatis, junioribus tomentosis, demum glabris, subtus ad venis pubescentibus; corymbis 3 ad 5-floris glabris, eglandulosis; stylis 4–5, fructibus maximis globosis rubris.*

CRATÆGUS *æstivalis.*—TORREY and GRAY, Flor., i. p. 468.
CRATÆGUS *elliptica.*—ELLIOTT, Sketch., i. p. 549.
CRATÆGUS *opaca.*—HOOK and ARNOTT, Compan. Botan. Magaz., vol. 1. p. 25.
Mespilus æstivalis.—WALTER, Flor. Caroliniana, p. 148.

This is another arborescent species of Hawthorn confined to the Southern States of the Union, growing along the low, wet banks of rivers and ponds, from South Carolina and Georgia to Florida: it grows also in Louisiana and Arkansas. In Florida, it is already in flower in the early part of the month of March, and presents a very unusual appearance, as yet nearly destitute of leaves, or presenting only their unfolding silky buds. The flowers are nearly as large as apple-blossoms, and pure white. It becomes at length a tree of twenty or thirty feet in height, branching from the base. The leaves are elliptical or oblong wedge-shaped, on the infertile branches often obovate, on short petioles, toward the summit sinuately toothed, angled, or irregularly crenate, rarely three-lobed or cleft, quite whitely tomentose when young before expansion, at length glabrous, but clothed along the veins beneath with a brownish pubescence. The corymbs are 3 to 5-flowered, and smooth. The divisions of the calyx are short, triangular, smooth, and without glands; the styles are four or five. The fruit, which becomes red, is very large and round, ripening in May or June, and is one-half or three-fourths of an inch in diameter, juicy, fragrant, of an agreeable subacid taste, and is much esteemed for tarts, jellies, and other articles of the dessert.

Other species of Hawthorn, indigenous to the United States, might be adduced as attaining the size of trees from ten to twenty-five feet in height; but, as we have little or no notice of their use and economy, we shall at present omit them.

CHERRY TREE.

Natural Order, ROSACEÆ, (*suborder*, AMYGDALEÆ, Jussieu.) *Linnæan Classification*, ICOSANDRIA, MONOGYNIA.

CERASUS. (JUSSIEU.)

Calyx urceolate-hemispherical; the border five-cleft, deciduous. *Petals* spreading. *Stamens* fifteen to thirty. *Ovary* glabrous, with two collateral pendulous ovules. *Drupe* globose, fleshy, destitute of bloom; the *nut* hard and bony, mostly globose and even.

Trees or shrubs chiefly of the temperate parts of Europe and North America, forming several natural sections. Leaves serrated, deciduous or sempervirent.

§ I. *Flowers corymbose or clustered. Leaves deciduous.* True CHERRIES.

SOFT-LEAVED CHERRY.

CERASUS MOLLIS. *Foliis oblongis ovatisve plerisque obtusis serrulatis subtus tomentoso-pubescentibus, corymbis racemosis 5 ad 6-floris tomentosis, laciniis calycinis obtusis reflexis tubo pubescente brevioribus, drupa ovoidea.*
CERASUS *mollis.*—DOUGLAS, in Hooker, Flor. Bor. Am., vol. i. p. 169.
TORREY and GRAY, Flor. N. Amer., i. p. 410.

THIS species of Cherry, growing twelve to twenty-five feet high, is confined to the Oregon Territory, and particularly to the borders of the Oregon River and its tributaries as far

Cerasus mollis.
Cerisier à feuilles molles.

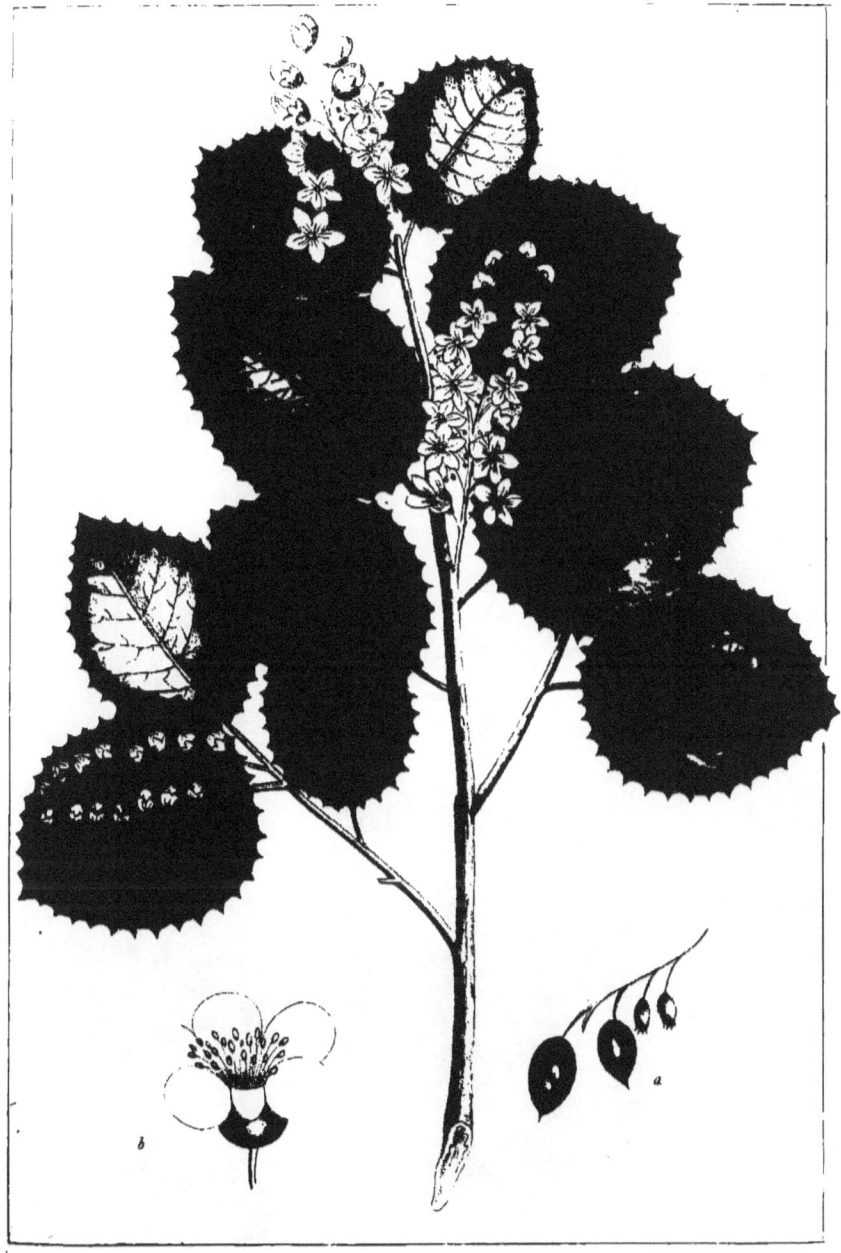

Cerasus Ilicifolia.

Holly leaved Cherry.   Cerisier à feuilles de Houx.

as its sources. We met with it in thickets on hills, near the Wahlamet, flowering about the month of May. The young branches are dark gray and somewhat downy. The leaves are softly downy beneath, on short petioles, oblong, or oblong-ovate, mostly obtuse, sometimes acute, minutely serrulate, two to two and a half inches long by about an inch in width; stipules small and deeply ciliate, as well as the bractes. Flowers small and white, the petals rounded and concave. Segments of the calyx ovate, short, and obtuse. Stigma clavate, petioles and calyx tomentose. Fruit ovate, astringent, and unpleasant.

PLATE XLVI.

A branch of the natural size, with young fruit. a. The flower.

RED or NORTHERN CHERRY. (*Cerasus Pennsylvanica*, TOR. and GRAY. *C. borealis*, MICH. and MICH., Sylva, p. 152.) According to Macmin, of West Chester, this tree in the Beech woods of Tioga county, Pennsylvania, attains the height of sixty feet, with a diameter of eighteen inches.

---

§ II. *Flowers in racemes, axillary. Leaves semperrirent or persistent.*—LAURO-CERASUS, Tourn., Decand., *Laurel Cherries.*

HOLLY-LEAVED CHERRY.

CERASUS ILICIFOLIA. *Foliis lato-ovalibus subcordatis brevi-petiolatis spinosa-sinuato-dentatis reticulatis coriaceis nitidis, racemis erectis foliis subæquantibus, drupa nigra ovoidea acuminata.*

CERASUS *ilicifolia.*—NUTT., in TORR. and GRAY, Flora N. Amer., vol. i. p. 411. HOOK and ARNOTT, Bot. Beechy, Suppl., p. 340, t. 83.

HOLLY-LEAVED CHERRY.

This is a small tree of Upper California, round Santa Barbara attaining the height of twelve to twenty feet, and chiefly affecting dry and elevated hill-sides within the mountain range. The bark is gray and somewhat rough; the wood is reddish, tough, and close-grained. The leaves, which are rigid, shining, and evergreen, look entirely like those of the Holly; they are broadly oval, pointed, somewhat heart-shaped at the base, very smooth and shining above and elegantly reticulated, often undulated, and with sharp pungent serratures. The racemes of flowers are erect, somewhat crowded; the flowers white and small, on short petioles; the petals rounded and short; the calyx hemispherical, with short triangular teeth. The stamens seated near to the summit of the calyx; the stigma simple and obtuse.

This tree, from its remarkable and elegant appearance, is well worth cultivating as an ornament, and in its qualities ranks with the true Laurels. The fruit is rather large, dark purple, bitter, and astringent.

PLATE XLVII.

A branch of the natural size. a. The cherry. b. The flower.

The LAUREL, (*Prunus Lauro-Cerasus*,) now so generally cultivated in Europe, was brought from Asia Minor. Lucullus, after conquering the King of Pontus, with whom the Romans had warred for forty years, among his other trophies, brought the Cherry from the fields of Cerasonte, and, in transplanting it into Italy, secured a monument of his triumph far more durable than that which the senate and the people decreed him. The Laurel, transplanted at first from Trebizond to Constantinople, had not so brilliant a destiny; an envoy from the Emperor of Germany, David Ungnad, whose name is now scarcely known, two hundred and sixty-two years ago brought a living plant to Clusius, at Vienna. The name of *Lauro-Cerasus* was given to it by Belon, who had seen it in its native country, from its leaves being like those of the Bay and its fruit similar to cherries.

The leaves afford by distillation a liquor which proves a violent poison to men and animals. According to Duhamel, a spoonful of this water given to a dog killed him instantaneously. Various experiments and accidents tend to confirm the fact of the powerfully-poisonous nature of Laurel water. Fontana found that a single drop of the essential oil of this plant, applied to a wound on a dog, proved equally as fatal as the venom of the viper, and was attended with the same symptoms.

The emanations from the Laurel being, in fact, the diluted but volatile prussic or hydrocyanic acid, are not without their inconveniences; for, after reposing beneath its shade on a warm day, a headache and tendency to vomit are said sometimes to occur. Considerable use was formerly made of Laurel-water for the sake of the Bitter Almond flavor which it communicates to various articles of the dessert, but from its dangerous effects it is now but little used.

The effect of this poison is so extremely rapid and violent, attacking the very seat of vitality, the nervous system, that no remedies have any time to operate. In the hand of the skilful physician, however, this volatile poison proves sometimes a powerful remedy.

ALMOND CHERRY, (*Cerasus Caroliniana*, MICH., Flor., vol. i. p. 285. Wild Orange Tree, MICH., Sylva, vol. ii. pl. 89.) This elegant tree, nearly allied to *C. Lusitanica*, appears to be common along the banks of the Mississippi from New Orleans to Natchez. It is also indigenous to South Carolina, Florida, and Arkansas. It forms a fine evergreen tree forty to fifty feet high, flowering in March and April. The leaves, according to Elliott, are very poisonous, frequently destroying cattle that are tempted to browse on them early in the spring. It is known to the French inhabitants of Louisiana by the same name as the Laurel of Europe, *Laurier-Amand.*

The fruit of this species is a small, black, bitter cherry, with very little pulp and a shell so thin as to crack between the fingers. A second (*C. Occidentalis*) and probably a third species of this section from St. Domingo, in the collections of Poiteau, has the same thin, fragile shell. These seem to form a separate genus from the true Cherries, no less than from the Laurels, and may be called LEPTOCARYA, in consideration of the thin and fragile, merely cartilaginous, shell of the drupe. In this respect the drupe affords a much more important distinction than that which exists between *Prunus* and *Cerasus.*

Prunus Americana.

Wild Plum — Prunier d'Amerique.

PLUM TREE.

Natural Order, ROSACEÆ, (*suborder*, AMYGDALEÆ, JUSS.) *Linnæan Classification*, ICOSANDRIA, MONOGYNIA.

PRUNUS. (TOURNEFORT.)

Calyx urceolate-hemispherical, the border 5-cleft, deciduous. *Petals* spreading. *Stamens* fifteen to thirty. *Ovary* glabrous, with two collateral, pendulous ovules. *Drupe* ovate or oval, fleshy, glabrous, usually covered with a bloom; nut hard and bony, more or less compressed, acute and even, the margins partly grooved.

Trees or shrubs of temperate climates in the Northern hemisphere, with the leaves serrated, convolute in vernation, (or before expansion.) The flowers earlier than the leaves, with the pedicels in umbellate clusters.

WILD PLUM.

PRUNUS AMERICANA. *Arborescens, ramis spinescentibus, foliis ovato-oblongis vel obovatis argute serratis acuminatis basi cuneatis, subtus venosis demum glabris, petiolis sub-biglandulosis, umbellis sessilibus paucifloris, (2-5,) fructibus ovalibus.*

P. AMERICANA.—MARSHALL, Arbust., p. 111. DARLINGTON, Flora Cest., p. 287, and in Annal. Lyceum, N. York, vol. iii. p. 87, t. 1. TORREY and GRAY, Flor. N. Amer., vol. i. p. 407.
P. NIGRA.—AITON, Kew., (ed. 1,) vol. ii. p. 165. Bot. Mag., t. 1117. PURSH, Flor. Am., vol. i. p. 331. WILLD., Sp. pl., vol. iv. p. 993.
P. HIEMALIS.—ELLIOTT, Sk., vol. i. p. 542.

CERASUS NIGRA, (*Loisel.*)—SERINGE, in DECAND., Prod., vol. ii. p. 538. HOOK., Flor. Bor. Am., vol. i. p. 167.

FEW plants in North America have a more extensive range than this species of Plum: it is met with from the Saskatchawan toward Hudson's Bay, and through all the intermediate country to Georgia, Louisiana, and Texas. In the western part of the State of New York it is very common, and, in some instances, (as it appeared to me in 1810,) it has been cultivated by the aborigines around their dwellings in the same manner as the Chickasaw Plum. When truly wild, it seems to affect the banks of streams and rich bottom-lands. In New Jersey, near Franklin Furnace, (Sussex county,) I have observed trees twenty to thirty feet high, and with trunks from six to fourteen inches in diameter. The ordinary height, however, is from fifteen to twenty feet. The wood is hard and of a reddish color, like that of the Wild Cherry, (*Prunus serotina*.) The fruit, when mature, which is in the month of August, is from half an inch to an inch in diameter, in some instances almost wholly yellow, but commonly vermilion-red on one side, wholly red, or a mixture of both colors, and in all the varieties covered more or less with a very evident bloom. When ripe, it contains a very sweet, thin pulp, with the disadvantage however of having a thick, bitterish, acerb skin; but by cultivation it is considerably improved, and the fruit is sometimes, as Dr. Darlington remarks, as large as a common apricot. In Upper Canada, where it was formerly cultivated, I have seen as many as twelve distinct varieties in the same orchard. It is also free from the attacks of the insects which have proved so fatal to nearly all the cultivated Plums.

The stem spreads out into a roundish head, with many rigid and somewhat thorny branches. The leaves are oblong-ovate and sometimes obovate, almost always narrowed below, with a distinct abrupt point or acumination, sharply serrated, strongly veined, and more or less pubescent beneath. The pedicels are smooth, two to five together, in clusters. Calyx pubescent, the segments lance-linear, serrulated at the apex; the petals oval or obovate, and rounded.

PLATE XLVIII.

A branch of the natural size in fruit.   a. A cluster of flowers.

CRAB APPLE.

Natural Order, ROSACEÆ, (*suborder*, POMEÆ, JUSS.) *Linnæan Classification*, ICOSANDRIA, PENTAGYNIA.

PYRUS. (LINN.)

Calyx-tube urceolate, adnate to the fleshy ovary, from which it is inseparable, with the border 5-lobed. *Petals* five, roundish, concave, on short claws. *Styles* usually five or less, distinct or conjoined at the base. *Pome* (or apple) fleshy, closed, internally 5-celled, the cells cartilaginous and 2-seeded. The seeds with a chartaceous coat.

Trees or shrubs (in the present section) with entire or palmately-lobed, serrated leaves. Flowers in terminal flattish clusters or corymbs. Fruit edible when not too acerb or astringent.

RIVER CRAB APPLE.

PYRUS RIVULARIS. *Foliis ovatis acutis indivisis junioribus trilobatis incisis argute serratis subtus pubescentibus, stylis (3–4) basi coalitis glabris, fructibus perparvis subglobosis vix umbilicatis, lobis calicinis demum deciduis.*

PYRUS *rivularis*.—DOUGLAS, in HOOK. Flor. Bor. Am., vol. i. p. 303, t. 68.

TORREY and GRAY, Flor. N. Am., vol. i. p. 471.

PYRUS *diversifolia*.—BONGARD, Veget. Sitka., l. c. p. 133.

This elegant species of Pyrus is common throughout all the lower or maritime portion of the Oregon Territory, and it uni-

Pyrus rivularis.

River Crab — Poirier rivulaire.

formly affects the shade of rich, alluvial forests near the lesser streams and ponds. It becomes a tree about the size of the Siberian Crab, to which it has a close affinity, and grows from fifteen to twenty-five feet in height, producing a hard wood, capable of receiving a high polish, and is employed by the natives for making wedges. The fruit grows in clusters, and is small and purple, scarcely the size of a cherry, of an agreeable flavor, like that of some of our Haws: it has nothing of the acerbity or acidity of the Common Crab, but is sweetish and subacid when ripe. The natives near the sea employ it, as they do many more berries of the country, for food, being all too indolent to cultivate the earth for any purpose whatever.

It extends, in all probability, from Upper California to the Russian possessions in the North, as far as latitude 57°. Menzies appears to have been its first discoverer, on what was then vaguely termed the northwest coast.

The leaves, which appear with the flowers, are ovate, obtuse or acute, entire, and more or less serrated, pubescent beneath, villous in the bud, at length nearly smooth; the later-produced leaves are more or less incisely lobed, sometimes distinctly three-lobed, the middle lobe incise and sharply serrated. The flowers conspicuous, white or tinged with red, in terminal corymbs, with the calyx and peduncles villous, or tomentose, at other times with the exterior of the calyx smooth. The petals oval. The germ is pear-shaped, with three or four styles. Apples very small, dark purple, almost black when ripe, and somewhat translucent, globose-ovoid, scarcely umbilicate at base, and with the summit naked, the calyx, as in the Siberian Crab, being deciduous. Seeds like those of the apple, and two in a cell, as usual.

I think it probable that the plants with "smooth pedicels and with the calyx externally smooth" ought to constitute a distinct variety, which may be termed *Pyrus rivularis β levipes.* In these the pedicels are also glandular.

What this plant may become by cultivation cannot yet be determined. The Siberian Crab, (now so ornamental and generally cultivated,) which also affects the alluvial borders of streams and rivers round Lake Baikal, and in Daouria, according to Pallas, in its native soil only attains the height of three or four feet, with a trunk about as thick as a man's arm, and full of tortuous branches. The berries, also, in Pallas's figure, (Flora Rossica, vol. i. tab. 10,) are not so large as ordinary peas, and pyriform or attenuate at the base like a pear. All this tribe of plants, so eminently serviceable both for ornament and use, deserve cultivation in a pre-eminent degree; and the present species has also the advantage of being perfectly hardy in all temperate and even cold climates, as it stretches along the coast nearly to the vicinity of Eastern Siberia.

All the plants of this section of *Pyrus* are natives of Temperate Europe and Northern Asia.

PLATE XLIX.

A branch of the natural size. a. *The apple.*

NARROW-LEAVED CRAB APPLE, (*Pyrus angustifolia*, AITON.) This appears to be scarcely more than a variety of the *Pyrus coronaria;* distinguishable, indeed, by its narrower leaves, usually entire, which are often acute below; but, as the styles are neither perfectly distinct nor constantly glabrous, and that the young leaves are also pubescent, no sufficient distinction remains. The fruit is likewise wholly similar.

Pyrus Americana.

American Mountain Ash. Sorbier d' Am

MOUNTAIN ASH.

§ III. *Leaves pinnate or pinnatifid; styles two to five, distinct; pome globose or turbinate; pulpy.*—SORBUS, Linn.

AMERICAN MOUNTAIN ASH.

PYRUS AMERICANA. *Foliis pinnatis glabris, foliolis oblongo-lanceolatis acuminatis inciso-serratis, serraturis setaceo-mucronatis, cymis compositis multifloris, fructibus globosis.*—DECAND., Prod., vol. ii. p. 637. TORREY and GRAY, Flor. N. Am., vol. i. p. 472.
SORBUS AMERICANA.—WILLD., Enum., vol. i. p. 520. PURSH, Flor., vol. i. p. 341.
SORBUS AUCUPARIA.—β MICH., Flor. Bor. Am., vol. i. p. 290.

THE Mountain Ash, or Rowan Tree, of North America, is met with sparingly in shady, moist woods in mountainous situations, from Labrador and even Greenland, throughout the New England States, New York, Pennsylvania; and the variety *microcarpa*, with smaller berries, extends to the high mountains of Virginia and North Carolina.

It forms a small tree of great beauty, remarkable for its elegant feathered foliage, in May and June clad with its white and fragrant blossoms, and to the close of the year, even into winter, decorated with its large clusters of bright berries, which afford a favorite repast for thrushes and other frugivorous birds, on their annual round to more genial climates, or during their hibernal residence:—

"Sanguineisque inculta rubent aviaria baccis."—VIRGIL.

The European species, which differs very little from the present, becomes, in the North of England, Scotland, and Wales, a tree of considerable size, so as occasionally to be sawn into planks and boards. It attains the height of twenty-five to thirty feet, with a diameter of two feet; and a tree in Scotland, in Forfarshire, at Old Montrose, sixty-five years old, is fifty feet high, with a diameter of two feet ten inches. The wood is said to be hard and durable, fit for economical purposes, such as mill-work, screws for presses, spokes for wheels, &c. In ancient times it was also esteemed for bows next to the Yew. The berries, dried and reduced to powder, have even been made into bread; and an ardent spirit may be distilled from them of a fine flavor, but in small quantity. Though acid and somewhat astringent, they are accounted wholesome, and, in the Highlands of Scotland, are often eaten when perfectly ripe; in the cold and sterile climate of Kamtschatka, according to Gmelin, they are used for the same purposes.

This tree was formerly held sacred, and in the North of England it is called the Witch-Hazel. In Wales, it was formerly planted in the churchyard as commonly as the Yew, and, on a certain day of the year, everybody religiously wore a cross made of the wood, as a charm against fascinations and evil spirits!

The American species scarcely forms so large a tree as that of Europe, attaining only the height of fifteen to twenty feet, and the leaves are very smooth, except before their complete expansion; the leaflets are about from thirteen to fifteen, oblong-lanceolate, acuminated, with sharp and deep mucronate serratures. The cymes or flower-clusters are large and compound, and the fruit, like that of the European species, is of a bright light scarlet. The berries of the variety *microcarpa* are also of the same color, but smaller. The seeds, two in a cell, appear to have the same cartilaginous coat as in the apple.

PLATE L.

A branch of the natural size. a. A cluster of flowers. b. A flower enlarged.

CERCOCARPUS.*

(HUMB., BONPL., and KUNTH.)

Natural Order, ROSACEÆ, (Juss.) (Sub-tribe, *Cercocarpeæ*.) *Linnæan Classification*, ICOSANDRIA, MONOGYNIA.

Tube of the *calyx* cylindrical, elongated, the lower part persistent, the border hemispherical, 5-lobed, deciduous. *Petals* none. *Stamens* many, seated on the border of the calyx. *Ovary* solitary; *style* terminal, filiform, and villous. *Achenium* narrow, coriaceous, caudate, with the long persistent and enlarging plumose style. *Seed* linear.

Shrubs or small trees, with alternate straight-veined, coriaceous, serrate or entire leaves on short petioles. Stipules small, adnate to the base of the petiole. Flowers small, white, axillary or terminating short branchlets, mostly clustered.

---

* The name derived from κερκος, a tail, and καρπος, a fruit, in reference to the character of the fruit.

FEATHER BUSH.

CERCOCARPUS LEDIFOLIUS. *Foliis crebris perennantibus lanceolatis integris demum glabris subtus tomentosis margine revolutis; floribus sessilibus paucis fasciculatis; cauda carpelorum longissimum tortuosum.*—NUTTALL, in TORREY and GRAY, Flor., Am., i. p. 427. HOOKER, Ic. pl., tab. 324, (ined.)

WE first observed this curious small tree in the Rocky Mountain range, on the lofty hills of Bear River of Timpanogos, near the celebrated "Beer Springs," which abound with carbonic acid. We saw it afterward in the central chain, on either side Thornberg's Ravine, toward the summits of the highest ridges, to which, by its enduring and dark verdure, it contributed to give a wild and gloomy robing, contrasted by the glittering white of the impending cliffs of gneiss near which it grew. On the summits of the Beer Spring hills it formed extensive thickets, each tree spreading out many branches at a few feet from the ground with considerable regularity, almost in the manner of a Peach Tree. The stem was in some trees about a foot in diameter, and the greatest height of the plant did not exceed fifteen feet. It had much the appearance of a stunted Olive Tree, and was bitterish to the taste.

The wood is hard, tough, whitish, and very close-grained, somewhat resembling that of the Birch. It appeared to be of slow growth and sempervirent; the bark smooth and whitish, the branchlets full of circular cicatrices, and the leaves clustered at the extremities of the twigs. The leaves are at length nearly smooth, at first hairy, with a short pubescence, beneath always softly villous, with brownish curled hairs; their form is lanceolate, about one and a quarter inches long and three or four lines wide, the border entire and revolute; beneath the hairs on the under side we see the usual straight nerves. The older leaves and other parts of the plant exude in small quantities an aro-

Cercocarpus ledifolius.

matic resin, having the scent of that found on some species of Birch, (or *Betula*.) The flowers are small and white, produced at the extremities of the twigs, and are succeeded by the fruit, which forms one of the most remarkable and singular characters of the genus; these have a strong resemblance to the seeds of the Geranium, each small cylindric carpel sending out a long, plumose, tortuous tail, nearly two inches in length, covered with yellowish-white silky hairs, which, appearing simultaneously all over the bush, give it a most remarkable and uncommon appearance. It seemed to prefer poor dry soils, and would bear the climate of Europe or the northern parts of the United States very well, from the alpine situations in which we uniformly saw it. It is somewhat astringent to the taste, and agreeably though not powerfully aromatic.

PLATE LI.

A branch of the natural size, with its fruit.   *a. The flower.*   *b. The fruit.*

WEST INDIA DOGWOOD.

Natural Order, LEGUMINOSÆ. *Linnæan Classification*, DIADELPHIA, DECANDRIA.

PISCIDIA.* (LINN.)

Calyx campanulate, 5-toothed. *Corolla* papilionaceous, with the keel obtuse. *Stamens* monadelphous, with the tenth free at the base. *Style* filiform, glabrous. *Legume* pedicellated, linear, with four broad longitudinal wings, the seeds separated by interruptions in the pod. The *seeds* oval and compressed, with a lateral hylum; embryo curved; cotyledons thick and elliptic; the radicle inflected.—West India trees, with deciduous, unequally-pinnated leaves, produced after the development of the flowers.

JAMAICA DOGWOOD.

PISCIDIA ERYTHRINA. *Foliolis ovatis, leguminis stipite calyce multo longiore, alis interruptis.*
PISCIDIA *Erythrina.*—LINN., Sp. pl. JACQ., Amer., p. 206. SWARTZ, Obs., p. 277. MACFADYEN, Flora of Jamaica, vol. i. p. 258.
Ichthyomethia foliis pinnatis ovatis, racemis terminalibus, siliquis quadrialatis.—BROWNE, Jamaica, p. 296.
Coral arbor polyphylla non spinosa, fraxini folio, siliqua alis foliaceis exstantibus rotæ molendinariæ fluviatilis acuta.—SLOANE, Jam., vol. ii. p. 32, tab. 176, figs. 4, 5. LAMARCK, Illust., tab. 605, fig. A.
Pseudo-acacia, siliquis alatis.—PLUMIER, Icon., 229, tab. 233, fig. 2.

---

* The name from *piscis*, a fish, in allusion to its employment as a fish-poison.

JAMAICA DOGWOOD.

The Jamaica Dogwood is a native of the Antilles as well as of the neighboring continent of America, having been observed by Humboldt and Bonpland in the mountainous places in New Spain, between Acapulco and Mazatlan, and we have now to record it as a native of Key West, in East Florida, where it was collected by Dr. Blodgett. It becomes a tree of about twenty to twenty-five feet in height, not remarkable for the elegance of its form, the branches being straggling, but yet beautiful in the season of flowering, which is about April, when, with blossoms similar to our favorite White Locust, (*Robinia pseudo-acacia,*) the whole summit of the tree is profusely loaded; they come out some time before the leaves, in numerous panicles or spreading clusters, of a whitish color, mixed with purple; the uppermost petal or vexillum in the centre tinged with green. The vexillum, externally, as well as the calyx, is covered with a silky pubescence. The leaves are unequally pinnate, with about five leaflets, which are either broad-ovate or obovate, and slightly acuminate, entire, and beneath, as well as the footstalk, more or less pubescent, particularly when young. The pod is large, stipitate, and villous, with four broad undulated longitudinal wings.

In Jamaica, this is esteemed one of the best timber trees in the island; the wood is heavy, hard, and resinous, coarse, cross-grained, and of a light brown color; it is very durable either in or out of water. It makes excellent piles for wharves; and the stakes soon form, in the tropical countries it inhabits, a good live fence. The bark of the trunk is very astringent: it cures the mange in dogs, and would probably answer well for the tanning of leather: it is best known, however, for its effects as a fish-poison, for which purpose it is pounded and mixed with the water in some deep part of a river or creek, when the water soon acquires a reddish shade, and in a few minutes the fish begin to rise to the surface, where they float, as if they were dead; the larger ones, however, recover, but the smaller fry are destroyed. The tincture of the bark, indeed, is found to be an intense narcotic, and has been employed beneficially to relieve the pain produced by carious teeth. Jacquin observes that this quality of intoxicating fish is found in many other American plants. *Tephrosia toxicaria* of South America and *T. piscatoria* of India and the South Sea Islands, both plants of the same family with the present, likewise possess the faculty of intoxicating fish.

PLATE LII.

A branch of the natural size. a. *The flowers and young pods.* b. *The more perfect pod.*

Acacia latisiliqua

Broad-podded Acacia / Acacia à large silique

ACACIA.*

(NECKER, WILLD.)

Natural Order, LEGUMINOSÆ. *Linnæan Classification*, POLY-
GAMIA, MONŒCIA.

Flowers POLYGAMOUS, perfect and staminiferous.— *Calyx* 4 to 5-toothed. *Petals* four to five, distinct, or united into a monopetalous, 4 to 5-cleft corolla. *Stamens*, from eight or ten to two hundred. *Legume* without interruptions between the seeds, dry, (without pulp,) and 1-valved.

These are trees and shrubs principally of warm or mild climates, with or without stipular or scattered spines. The leaves are usually small and variously pinnated: sometimes (particularly in the New Holland species) the true leaves in the adult are abortive, and the simple leafy petioles, called phyllodes, alone supply their place. Flowers often yellow, more rarely white or red, disposed in spherical heads or in spikes.

BROAD-PODDED ACACIA.

ACACIA LATISILIQUA. *Inermis glabra, pinnis 5-jugis, foliolis 10–15-jugis ellipticis obtusis, stipulis bracteiformibus dimidiato-cordatis, capitulis pedunculatis aggregatis in paniculam terminalem subdispositis, legumine longe stipitata, plana, utrinque acuta.*—DECAND., Prod., vol. ii. p. 467.
ACACIA LATISILIQUA. *Inermis, foliis bipinnatis partialibus quinquejugis, ramis flexuosis, gemmis globosis.*—LINN., Sp. pl. PERSOON, Synops.,

---

* An ancient Greek name, from αχαζω, *to point*, or *sharpen*, many of the species being thorny.

vol. ii. p. 265. WILLD., Sp., vol. iv. p. 1067. MACFADYEN, Flor. Jam., vol. i. p. 318.

Acacia non spinosa, siliquis latis compressis, flore albo.—PLUMIER, (Ed. Burm.,) t. 6.

THIS species, like many others of the genus, remarkable by its light, waving, feather-like foliage, is, according to Dr. Blodgett, rare at Key West, where it becomes a very large and spreading tree, flowering in the month of May. It is also a native of the West Indies and the warmer parts of the neighboring continent, where it was found by Plumier and Aublet. According to Macfadyen, it is a cultivated plant in Jamaica. It bears a great resemblance to the Acacia figured by Catesby, tab. 42, which is quoted as *A. glauca,* though by no means the same plant as Plate 36 of Trew, which latter is the species most commonly cultivated under that name.

The wood of this Acacia is said to be white, hard, and close-grained. The trunk, as described by Catesby, attains a diameter of three feet, and is accounted an excellent wood, next to the mahogany of Jamaica, and is the best to be found in the Bahama Islands. For curious cabinet-work it excels mahogany in its variable shining tints, which appear like watered satin. Several species of the genus afford very hard and durable wood.

The small branches in this species are gray, slender, and somewhat zigzag. The leaves are bipinnate, on main petioles, a little more than an inch long; between the first pair of pinnules is usually seen on the petiole a projecting though sometimes merely a depressed gland; the next pairs are without glands to the summit of the leaf-stalk, where there is then another depressed gland. The pinnules vary in our plant from two to four pair; (we have not seen five.) The leaflets of the pinnule are oblong-elliptic, nearly smooth, obtuse, somewhat oblique, and rounded at base, in from eight to fifteen or sixteen pairs. From the axils of the two or three uppermost leaves come out simple or aggregated peduncles, usually by threes, above, running together so as to form a small, sparse-flowered panicle, with each of the clusters subtended by rather large, deciduous, amplexicaule, semicordate and acuminate smooth bractes, which resemble stipules. The flowers are disposed in spherical, rather small heads, on peduncles about three-quarters of an inch long: they appear white from the color of the long, tortuous, hair-like stamens. The calyx is canescent, with a close pubescence, and five-cleft at the summit. The corolla is deeply five-parted, and of a purplish brown, with oblong-lanceolate divisions. The stamens are ten or more, with very long filaments, and very small whitish rounded anthers. The legume (according to Dr. Blodgett) is four or five inches long, flat, thin, many-seeded, and an inch or more in breadth.

PLATE LIII.

A small branch of the natural size. a. The flower somewhat enlarged.

INGA.*

(PLUMIER, WILLD.)

Natural Order, LEGUMINOSÆ.   *Linnæan Classification*, POLY-
GAMIA, MONŒCIA.

Flowers POLYGAMOUS, perfect, and male.—*Calyx* 5-toothed. *Corolla* monopetalous, tubular-funnel-formed, exceeding the calyx in length, with the border regular and 4 or 5-cleft. *Stamina* numerous, exserted, (10 to 200,) with the capillary filaments more or less united into a tube. *Legume* broadly linear, compressed, 1-celled. *Seeds* usually covered with pulp, more rarely with a pellicle or with farinaceous matter.

Shrubs or trees of warm or tropical climates, chiefly indigenous to India and America, usually unarmed. Flowers in spikes or globular heads, red or white, rarely yellow.

BLUNT-LEAVED INGA.

INGA UNGUIS-CATI.† *Spinis stipularibus rectis, foliis conjugato-geminatis, foliolis subrotundo-ellipticis subdimidiatis membranaceis glabris, glandula*

---

* An American name adopted by Plumier.

† The specific name of *unguis-cati* alludes to the short and rather concealed thorns with which this tree is provided. Browne calls it the *black-bead* shrub, and from others in Jamaica, according to Macfadyen, it receives the names of *Barbary Thorn* and *Nephritic Tree*.

Inga Unguis Cati.

Blunt leaved Inga.   Inga ongle de chat.

BLUNT-LEAVED INGA.

in dichotomia petioli glabri et inter foliola, florum capitulis globosis in racemum terminalem dispositis, legumine torto.—DECAND., Prod., vol. ii. p. 436.

MIMOSA *unguis-cati.*—LINN., Spec., 499. WILLD., Sp. pl., vol. iv. p. 1006. JACQUIN, Hort. Schoenbrunn, vol. ii. tab. 34. DESCOURT., ·Flor. Antil., vol. i. tab. 11. SWARTZ., Obs., p. 389. MACFADYEN, Flor. Jam., vol. i. p. 306.

Acacia quadrifolia, siliquis circinatis — PLUMIER, (Ed. Burman,) Icon. 4. PLUKEN., tab. 1, fig. 6.

Acacia arborea major spinosa, pinnis quatuor, siliquis varie intortis.— SLOANE, Hist. Jam., vol. ii. p. 56.

Mimosa fruticosa, foliis ovatis binato-binatis, seminibus atro-nitentibus.— BROWNE, Jamaic., p. 252.

This very singular-leaved tree, attaining about the height of from ten to twenty feet, is indigenous to many of the West India Islands, as well as to Cumana and Cayenne on the neighboring continent, where it was observed by Humboldt and Bonpland, and in the latter place by Aublet. This is also another of the Caribbean productions which extends to the limits of the United States, having been recently found in Key West by our friend Dr. Blodgett.

The wood is said to be yellow, the summit of the tree irregular, and the branches straggling. The smaller twigs are round and gray, inclining to brown, and covered with minute warts. The thorns are stipular, or come out at the junction of the leaf with the stem; they vary in size, but are always short, and in some of the twigs wholly absent. The leaves are bipinnate, only four in number, the leaflets on each pinule being only a single pair, sessile, obovate, very obtuse or subemarginate and rounded above, glabrous and of a thin texture, with widely-reticulated nerves; the petiole channelled above, with a hollow circular gland at the junction of the secondary petioles. Racemes terminal, thyrsoid, the pedicels long and fastigiate, almost like a corymb. Flowers greenish yellow and smooth, in globose heads. Calyx small, five-toothed. Corolla more than twice the length of the calyx, five-cleft toward the summit, the segments acute. Filaments numerous, slender, and capillary, yellow, three times the length of the corolla. Legume torulose, spirally twisted, of a reddish-purple colour; seeds five or six, black, shining, roundish, compressed, half covered with a white, fleshy, arillus-like pellicle.

This plant has the credit of being a sovereign remedy for nephritic complaints, for the stone and gravel, and also for obstructions of the liver. The bark is the part employed; and Barham states (in his account of Jamaica, where this tree grows) that in his time it was in such general use that it was rare to meet with a tree that had not been barked. The decoction, of a red color, is very astringent, and acts as a diuretic. It has also been employed externally as a lotion and injection, to remove the relaxation of the parts. Upon the whole, it would seem to be entitled to the notice of physicians, and deserves a further examination.

PLATE LIV.

A branch of the natural size. a. The flower somewhat enlarged.

GUADALOUPE INGA.

INGA GUADALUPENSIS. *Inermis, foliis conjugato-geminatis, foliolis obovatis subrhombeis obtusis venosis glaberrimis, glandula in dichotomia petioli glabri et inter foliola, capitulis globosis pedicellatis racemosis, legumine torto glabro.*—DECAND., Prod., vol. ii. p. 436.

MIMOSA GUADALUPENSIS. *Foliis bijugis foliolis ovalibus, obliquis subcoriaceis, capitulis corymbosis.*—PERSOON, Synops., vol. ii. p. 262.

THIS species also becomes a tree of twelve to twenty feet elevation at Key West, according to the observation of the same

Inga Guadalupenis

Guadaloupe Inga   Inga de la Guadeloupe.

gentleman who discovered the preceding. The specimen described by Persoon came from the island of Guadaloupe. Decandolle suspects that it may be a mere thornless variety of the preceding species, (*I. unguis-cati;*) but, from numerous specimens which we have inspected from Florida, there can remain very little doubt of its distinction as a peculiar species.

The spines appear to be wholly wanting; the bark of the branches is gray and rough with minute warts. The petioles are about three lines long, and of the same length with the partial ones; both are strongly grooved and distinctly articulated. The leaves are smooth and coriaceous, shining above, dull and paler beneath, delicately and reticulately veined, quite opaque from their thickness, cuneate-oblong or lanceolate-oblong, obtuse, and sometimes rounded at the apex, at other times rather acute and apiculated. A depressed gland at the summit of the petiole between the stalks, and also one less distinct between the pairs of leaflets. The flowers are axillary and long-pedunculate; they likewise terminate the branches in corymbose racemes. The heads of flowers are hemispherical, and appear to have been yellowish green. The calyx is campanulate, with acute and very distinct teeth; the corolla is monopetalous, more widely campanulate at the summit, twice as long as the calyx, with acute segments. The pods are dark purplish brown, much curved, three to four inches long, about half an inch wide, attenuated at the base, torulose and irregularly narrowed between the seeds, but not intercepted within. The seeds are deep black, somewhat compressed, and at one extremity half covered by a bright rose-red fleshy and lobed arillus.

PLATE LV.

A branch of the natural size. a. The ripe pod. b. The seed.

SCHÆFFERA.*

(JACQUIN.)

Natural Order, CELASTRINEÆ? *Linnæan Classification*, DIŒCIA, TETRANDRIA.

DIŒCIOUS.—*Calyx* small, 4-parted, persistent. *Petals* four, alternating with the sepals. *Stamina* four, opposite to the petals. *Ovarium* 2-celled. *Stigmas* two. *Berry* dry, bipartite, cells 1-seeded. *Seed* erect, plano-convex; albumen fleshy; embryo central, straight, and flat.

Trees of Tropical America, with alternate, entire, coriaceous leaves; stipules none; flowers several, axillary, small and pedicellated, white or green.

JAMAICA BOXWOOD.

SCHÆFFERA BUXIFOLIA. *Foliis lanceolato-ovatis basi attenuatis plerisque acutis ramulisque glabris, petalis viridis obtusis.*
SCHÆFFERIA FRUTESCENS, *buxifolia.* Foliis latius ovatis mucronatis. —DECAND., Prod., vol. ii. p. 41. LAM., Illust., t. 809.
Buxi folio majore acuminato, arbor baccifera, fructu minore croceo dipyreno. —SLOANE, Hist. Jamaica, vol. ii. p. 102, tab. 209, fig. 1.

ACCORDING to Dr. Blodgett, this plant, common at Key West and on the adjoining keys of East Florida, becomes a tree of

---

* Named in honor of James Christian Schæffer, of Ratisbon, author of several botanical works.

Schœffera buxifolia.

thirty feet in height, and is an article of export from the Bahama Islands, where it is valued at about forty dollars the ton. From Poiteau's "Herbarium," it appears to grow in the island of St. Domingo; it is also apparently identical with the Jamaica plant of Sloane. The wood is pale yellow, very close and fine-grained, and might easily be mistaken for that of the true Box, which name it bears in the Bahamas.

The twigs are slender and covered with a light gray bark. The leaves are very smooth and shining on the upper surface, with slender branching veins, lanceolate and very acute, yet on the lower part of the same specimen blunt or even emarginate; but they are always narrowed below. The male flowers (the only ones I have seen) are small, on very short peduncles, three or four together, with a rather minute calyx, and four broadish, green, oblong, obtuse petals. The stamens are usually four, shorter than the petals, sometimes more by the ingraftment of two peduncles. The stigmas are two, and short. The berries rather flattened and two-lobed, about the size of a grain of cubebs, dry, but with a thick integument, two-celled, two-seeded, and of a pale orange-yellow when ripe. Appearances of resin are visible on some of the buds, and the berries have rather an acrid bitter taste, something like that of tobacco; yet, notwithstanding their disagreeable taste, they are greedily devoured by birds.

The white flowers of *S. frutescens*, the *S. completa* of Swartz, and its humble stature, appear to distinguish it from our plant.

PLATE LVI.

A branch of the natural size. a. The male flower. b. The fruit.

CEANOTHUS.*

(LINN., in part.)

Natural Order, RHAMNEÆ, (Decand.) *Linnæan Classification,*
PENTANDRIA, MONOGYNIA.

Calyx campanulate, shortly 5-cleft, with the border deciduous. *Petals* five, cucullate, and arched, exserted, with long claws. *Stamens* exserted. *Disk* thickened at the margin surrounding the ovary. *Styles* three, united to the middle. *Fruit* dry and rigid, mostly 3-celled, obtusely triangular, seated on the persistent tube of the calyx, tricoccous, dehiscing by the inner sutures. *Seeds* obovate, even.

Shrubs or undershrubs, rarely small trees, of the temperate parts of America. Roots large and ligneous. Leaves alternate, ovate or elliptical, mostly serrate, sometimes entire, persistent or deciduous. Flowers white or blue, in umbel-like clusters, aggregated at the extremities of the branches into thyrsoid corymbs. The taste of the root and most other parts of the plant more or less astringent. One of the species was formerly employed as a succedaneum for tea, and hence the name of *"New Jersey Tea."*

---

* An ancient Greek name employed by Theophrastus for a plant now unknown.

Tree Ceanothus.

Ceanothus thyrsiflorus. *Ceanothe thyrsiflore.*

TREE CEANOTHUS.

CEANOTHUS THYRSIFLORUS. *Arborea, erecta; ramis angulatis, foliis ovato-oblongis, subellipticis, obtusis crassiusculis, glanduloso-serrulatis subglabris, subtus subvillosis; thyrsis oblongo-ovalibus densifloris corymbulis axillaribus terminalibusque, ramis floriferis foliosis; floribus azureis.*
CEANOTHUS *thyrsiflorus.*—ESCHOLTS, in Mem. Acad. St. Petersb., (1826.)
HOOKER, Flor. Bor. Am., vol. i. p. 125. HOOKER and ARNOTT, in Bot. Beechy, p. 136. TORREY and GRAY, vol. i. p. 266.

THOUGH several species of this elegant genus in California, Oregon, and along the northwest coast, become considerable shrubs, this is the only one which can be classed among trees. It was somewhat abundant on dry, gravelly hills in the vicinity of Monterey, where I arrived in the month of March, about the time that it was bursting into flower. My attention was called to it in the wood-pile, where considerable stems, at least as thick as a man's leg, lay consigned to the ignoble but still important use of firewood. The wood appeared hard, tough, of a reddish color, and it afforded a durable fuel. The branches were tortuous, spreading, and covered with a rough bark; the branchlets green and angular. Leaves nearly elliptic, the uppermost ovate-oblong, all glandularly serrulate; above smooth, beneath pubescent, particularly along the three strong nerves which traverse the leaf to the summit; the petioles very short; the upper branchlets terminating in thyrsoid panicles of deep blue and very elegant flowers, made up of numerous round, dense clusters, in small corymbs; the terminal mass oval, about three inches long by about an inch in width; the clusters are subtended by ovate, acuminate, broad, villous, and deciduous bractes. The calyx, petals, and peduncles, are of a deep sky-blue; the segments of the calyx ovate; the petals, as usual, unguiculate and exserted, as well as the stamens; the anthers are yellow. With the fruit I am wholly unacquainted.

As this is a hardy and very ornamental plant, it well deserves cultivation. The flowers appear early in the spring, and the whole summit of the tree appears of an intense blue.

The bark of the *Ceanothus azureus*, a plant allied to the present species, is esteemed in Mexico as a febrifuge.

PLATE LVII.

A branch of the natural size. a. The flower.

CEANOTHUS *macrocarpus*.—NUTT., in TORREY and GRAY. As this is not the plant of Willdenow, I take this opportunity of correcting the error, and propose to call it *Ceanothus megacarpus*.

PERSIMMON, (*Diospyrus Virginiana*.) β PUBESCENS. *Foliis subtus molliter pilosis.*

Of this remarkable variety, with the leaves softly pilose beneath, I have seen specimens from Louisiana, collected by Mr. Teinturier; and a very similar but less pubescent variety was found in Georgia by the late Dr. Baldwin, (according to specimens in the herbarium of the Academy of Natural Sciences in this place.)

Snake Wood.

Colubrina Americana.   *Bois de Couleuvre.*

COLUBRINA.

(RICHARD.)

Natural Order, RHAMNEÆ. *Linnæan Classification,* PENTANDRIA, MONOGYNIA.

Calyx spreading, 5-cleft; the tube hemispherical. *Petals* five, obovate, convolute. *Stamens* five, with ovate, 2-celled anthers. *Disk* fleshy, rather flat, slightly 5-angled. *Ovary* immersed in and adhering to the disk, 3-celled. *Style* trifid. *Stigmas* three. *Fruit* capsular, dehiscent, tricoccous, girt at the base by the adnate, permanent, entire tube of the calyx. *Seeds* furnished with a short stalk, the testa coriaceous, very smooth.

Trees or shrubs of Tropical America and Asia. Leaves alternate, with pinnate nerves, and reticulated with transverse veins. Flowers in short, axillary cymes.

SNAKE-WOOD.

COLUBRINA AMERICANA. *Foliis ovatis subacuminatis integris, subtus ramulis floribusque ferrugineo-villosis, floribus axillaribus corymboso adgregatis.*
CEANOTHUS *colubrinus.*—LAMARCK. DECAND., Prod., vol. ii. p. 31. PERSOON, Synops., vol. i. p. 244.
RHAMNUS *colubrinus.*—JACQUIN, Amer., 74, No. 2, Hort. Vindobon., vol. iii. tab. 50. VOGEL, Icon. rar., tab. 105. LINN., Syst., vol. i. p. 195.

SNAKE-WOOD.

RHAMNUS *arboreus, foliis obovatis venosis, capsulis sphæricis, inferne ad medietatem calyptratis.*—BROWNE, Jamaic., p. 172, No. 2.
RHAMNUS *ferrugineus.*—NUTT., in TORREY and GRAY, Flora N. Am., vol. i. p. 263, and Journ. Acad. Nat. Sc., Philad., vol. vii. p. 90.
Arbor baccifera indica, foliis majoribus splendentibus flore pentapetalo.— COMM., Hort., p. 475, tab. 90.

A FLOWERING specimen of this tree was collected at Key West, in East Florida, by Mr. Titian Peale. From this imperfect relic I conceived it to belong to a new species, which I hence called the ferruginous Buckthorn; but on comparing it more attentively with a fine specimen of *Rhamnus colubrinus*, collected in St. Domingo by Poiteau, I felt satisfied of their identity. It is indigenous to the islands of St. Martin, the Bahamas, Jamaica, St. Domingo, and Cuba, where, on the high mountains, it becomes a tree of twenty feet in height; but on the borders of the sea, among the brushwood, it seldom attains a greater height than that of six or seven feet. The branches spread out horizontally and are thickly covered with leaves. It is remarkable for the ferruginous down spread over the petioles and young leaves, as well as upon the peduncles and calyx of the flowers. The bark is smooth and blackish, but the younger branches are gray and downy. The leaves are alternate, oval, somewhat acuminately and abruptly pointed, entire, smooth and shining above, tomentose beneath when young, afterward only so on the nerves, three to four inches long by about two inches wide; the petioles from a quarter to half an inch long. The flowers are small, disposed in short, axillary corymbs, containing in each cluster about seven to ten. The calyx is villous and ferruginous, five-parted, the divisions ovate and somewhat acute; the petals, five in number, are narrow, linear-oblong, about the length of the divisions of the calyx, unguiculate, concave, and partly embracing the stamens, which are about the same length. Fleshy disk of the germ conspicuous, broadly five-lobed. The style is simple, terminating in three simple, obtuse stigmas. The fruit, nearly half-way embraced by the persistent base of the calyx, is a capsule of three lobes, with three valves and three elastic cells. The seeds are solitary, nearly round, and somewhat compressed, shining and black, remaining, often after the lapse of the capsule, attached to the base of the cells. With the wood of this tree or its economy I am unacquainted.

Another species of this genus, with smooth, elliptic, and somewhat acuminated leaves on longish petioles, occurs, according to La Sagra, in Cuba. In this also the small axillary umbels are very few-flowered, smooth, and pedicellated: this might be called *Colubrina glabra*.

PLATE LVIII.

A branch of the natural size. a. The umbel of flowers. b. The flower a little enlarged. c. The seed remaining attached to the receptacle.

BUCKTHORN.

Natural Order, RHAMNEÆ.   *Linnæan Classification*, PENTANDRIA, MONOGYNIA.

RHAMNUS.* (LINN.)

Calyx urceolate, with the border 4 or 5-cleft. *Petals* four or five, alternating with the calyx, entire, emarginate or 2-lobed, more or less convolute, sometimes wanting. *Torus* thin, lining the tube of the calyx. *Stamina* situated before the petals. *Ovary* free, and not immersed in the torus or disk, 2 to 4-celled. *Styles* two to four, distinct, or combined. *Fruit* drupaceous, containing two to four cartilaginous nuts.

The Buckthorns are all shrubs or small trees, with alternate and rarely opposite leaves, on short petioles, often pennately nerved. The flowers are small and greenish, usually in short axillary clusters or small corymbs.

CAROLINA BUCKTHORN.

RHAMNUS CAROLINIANUS, (WALTER, Flor. Carol., p. 101.) *Erectus, foliis ovali-oblongis integriusculis glabris, umbellis pedunculatis, floribus hermaphrodites, fructibus globosis.*—MICHAUX, Flor. Bor. Am., vol. i. p. 153. DECAND., Prod., vol. ii. p. 26.

RHAMNUS CAROLINIANUS. Erect, unarmed; leaves oval-oblong, obscurely serrate, nearly glabrous, (or rarely pubescent beneath;)

---

* From the Celtic *ram*, branching; and hence the Greek *ῥάμνος*.

Carolina Buckthorn.

CAROLINA BUCKTHORN.

umbels axillary; on peduncles much shorter than the petioles; flowers perfect, pentandrous, (sometimes tetrandrous;) petals minute, embracing the very short stamens; styles united to the summit; stigmas three; fruit globose, rather dry, 3 to 4-seeded.— TORREY and GRAY, Flora N. Amer., i. p. 262.

THIS fine Buckthorn, though usually a shrub in our Southern and Southwestern forests, on the borders of Palmetto Creek, Laurens county, in Georgia, the late Mr. Croom observed trees of this species thirty to forty feet high. In the forests of Arkansas, they attain the height of ordinary Peach or Apple Trees, and, congregated together, produce shady groves of considerable extent. The quality or uses of its wood remain a desideratum. The stems are, however, slender for their height, being not more than four to six inches in diameter.

The leaves are three to six inches long and one to two inches wide, oval-oblong and widening toward the summit, the extremity more or less briefly acuminate, the border slenderly serrulate, and sometimes irregularly waved; the lateral pennate veins are ten to twelve, and rather distant; the very young leaves before expansion are somewhat ferruginously villous. Umbels on stout pedicels, from 10 to 15-flowered. The calyx pubescent but not ferruginous, the segments lanceolate; petals two-lobed at the extremity. The fruit, black, as large as a small pea, is mostly three-seeded. Seeds black, plano-convex, without a groove.

This species begins to appear in North Carolina, and extends through Georgia to Florida. West of the Mississippi, it is abundant on the banks of the Arkansas, and Mr. Say collected it within the range of the Rocky Mountains.

In Bartram's Botanic Garden, at Kingsessing, where this species is perfectly hardy, it forms an elegant tree, and has attained the height of nearly twenty-five feet in twenty years. Colonel Carr, the late worthy proprietor of this interesting garden, tells me that for a considerable time the berries remain red, and are very ornamental; at length toward winter they turn black, and remain so for a long time, until some famished flock of robins falls upon and strips them nearly at once.

PLATE LIX.

A branch of the natural size.   a. The flower enlarged.   b. The berries.

PURSH'S BUCKTHORN.

RHAMNUS PURSHIANUS, (Decand.) *Inermis, erectus, foliis lato-ellipticis minute denticulato-serratis subtus pubescentibus nervis lateralibus obliquis lineatis, pedunculis axillaribus umbellatis floribusque pubescentibus, calyce 5-fido, petalis minutis cucullatis.*—HOOK., Flor. Bor. Am., vol. ii. p. 123, t. 43. DECAND., Prod., vol. ii. p. 25.
RHAMNUS *alnifolius.*—PURSH, Flor. Am. Sept., vol. i. p. 166, (non *L'Heritier.*)

THIS is another species of Buckthorn which becomes a tree of ten to twenty feet elevation, with a trunk of nine inches in diameter. It was discovered within the Rocky Mountain range, on the banks of Salmon River, by Captain Lewis; and it is of common occurrence on the borders of the Oregon, in the upland shady woods near the skirts of the prevailing Pine forests. Menzies also met with this tree near Nootka, on the northwest coast of America. It bears a strong resemblance to the *R. Carolinianus;* but the leaves are broader, shorter, and more decidedly serrated, and the berry is strongly three-lobed.

The branches are round, dark brown, and pubescent. The leaves are three to five inches long, petiolate, deciduous, but at length somewhat coriaceous, broadly elliptic, rounded or rarely somewhat acute at the base, obtuse or sometimes very shortly

CATHARTIC BUCKTHORN.

acuminate, the margin minutely serrulate, the young leaves pubescent at length, only so on the nerves beneath, the nerves in oblique lines; petioles pubescent. Stipules quickly deciduous, peduncles solitary, an inch or more long, umbellated; pedicels pubescent, elongated in the fruit. Calyx externally pubescent, six-cleft; the segments acute, internally carinate. Petals minute, cucullate, bifid at the apex, shorter than the calyx, very concave, and cucullate. Stamens opposite the petals and involved in them. Germ small, ovate. Style shorter than the germ; the stigma obtuse and three-lobed. Berry wider above, three-celled, three-seeded. The seed obovate, black, very shining, convex externally, internally with a central, elevated line at the base, at the hylum yellow.

The CATHARTIC BUCKTHORN (*Rhamnus catharticus*) appears to be a native of the Northern States of the Union, as it occurs in the wildest situations. The berries and syrup of this species have long been employed in medicine. The juice of the berries, in a dose of five or six drachms, proves a strong cathartic; but it is generally made into a syrup. The bark has also an emetic quality. The juice of the unripe berries with alum gives a yellow dye; that of the ripe fruit, concentrated by evaporation, and treated in the same manner with a solution of alum, gives a green paste,—the sap-green employed by painters,—and, from the manner in which it is prepared for sale, is called, in France, *vert de vessie*.

In New England, particularly in the vicinity of Boston, this species is much employed for useful and ornamental hedges, and, bearing well to be cut, growing thick, and remaining green till winter, it is strongly recommended for this useful purpose.

MANCHINEEL.

Natural Order, EUPHORBIACEÆ? *Linnæan Classification*, MONŒ-CIA, MONADELPHIA.

HIPPOMANE.* (LINN.)

MONŒCIOUS.—*Male* flowers with a subcampanulate, emarginate calyx and no corolla. A single columnar filament terminating in four anthers.—In the *fertile* flower there is a 3-leaved calyx and no corolla. *Style* very short. *Stigma* 6 or 7-cleft. *Fruit*, a drupe containing a six to seven or more celled nut; each cell with one seed; the cells indehiscent.

A large poisonous tree of Tropical America, with alternate, entire leaves; the male flowers clustered in interrupted, terminal spikes. The fruit solitary and sessile, resembling an apple.

MANCHINEEL.

HIPPOMANE MANCINELLA. *Foliis ovatis serratis.*—LINN., WILLD., Sp. pl. LAMARCK, Illust., t. 793. JACQ., Am., edit. pict., t. 238. AUBLET, Guian., vol. ii. p. 885.
Malus Americana, laurocerasi folio, venenata. Mancinello arbor seu Massinilia dicta.—COMMEL., Hort., vol. i. p. 131, t. 68.
Juglandi affinis arbor julifera, lactescens, venenata, pyrifolia, Mancanillo *Hispanis dicta.*—SLOANE, Jamaic. Hist., vol. ii. p. 3, t. 159.

---

* From ἵππος, a *horse*, and μανια, *madness*. The name, however, was applied by the Greeks to a very different plant which grew in Arcadia, said to render horses furious.

Mancanilla pyrifacie.—Plumier, Gen., p. 49, t. 3. MSS. vol. vi. t. 109. Catesby's Carol., vol. ii. p. 95, t. 95.

Arbor Americana Mancinello *dicta, fructu pomi venenato, nucleis septenis et pluribus, in ossiculo muricato, totidem loculis dispertito, inclusis.*—Pluken, Almag., p. 44. Phytog., tab. 142, fig. 4.

Hippomane arboreum lactescens, ramulis ternatis; petiolis glandulâ notatis; floribus spicatis, mixtis.—Browne, Jam., p. 351.

The Manchineel Tree attains a great size on the sea-coast in various parts of the West India Islands and the neighboring continent. It has also been found growing very common at Key West, in low places, where it attains the height of thirty to forty feet. It has much the aspect of a Pear Tree at a distance, while the fruit resembles in appearance and scent a small apple, and is produced in such abundance that the ground, when they fall, appears as if it were paved with them; they possess, however, very little pulp, being internally occupied by a deeply-grooved nut as large as a chestnut. No animal, except goats and macaws, chooses to feed on them; and they become dry, brown, and spongy, and as useless as they are deleterious. The wood, on the contrary, is in great esteem for tables, cabinets, and other articles of furniture, being close-grained, heavy, durable, finely variegated with brown, white, and shades of yellow, and susceptible of a high polish. Tables made of it almost resemble marble, and are equally smooth and shining. Great caution, however, is necessary in felling the tree; and, before they begin, it is the usual practice of the workmen, first to kindle a fire round the stem, by which means the milky sap becomes so much inspissated as not to follow the blows of the axe. They also take the further precaution to cover the face with a net of gauze, to prevent the access both of the juice and the particles of sap-wood, which might be deleterious.

All parts of the Manchineel Tree abound with a white, milky sap, which is very poisonous, and so caustic that a single drop received upon the back of the hand immediately produces the sensation of the touch of a coal of fire, and soon raises a watery blister. The Indians, according to Hawkins, used to poison their arrows with this juice, which retained its venom for a long time. Another and much more deadly poison was commonly used for this purpose, however, by the American savages of the warmer parts of America,—namely, the *warari*, chiefly obtained from the juice of the *Strychnos;* and this was distinguishable by producing the effect of *tetanus* or lockjaw, which, mostly fatal, was sometimes protracted for several days before producing death. It is reported that many of the Europeans who first landed in Surinam died suddenly from sleeping under this tree; and there may probably be some foundation in truth for such reports, when we take into consideration the volatile nature of the poisonous principle of these plants. As in the venomous species of Rhus or Sumach, also, while many individuals are affected by the poison, others, for no evident reason, can touch or handle these plants with impunity. Hence, though Jacquin asserts that he reposed under the shade of the Manchineel for the space of three hours without experiencing any inconvenience, it does not follow that it would be equally harmless to all who should hazard the experiment; and, with a laudable prudence, the inhabitants of Martinique formerly burned down whole woods of the Manchineel in order to clear their country of so dangerous a pest.

Catesby acknowledges that he was not sufficiently satisfied of its poisonous qualities "till, assisting in the cutting down a tree of this kind on *Andros* Island, I paid for my incredulity: some of the milky poisonous juice spirting in my eyes, I was two days totally deprived of sight, and my eyes and face much swelled, and felt a violent pricking pain the first twenty-four hours, which from that time abated gradually with the swelling, and went off without any application or remedy, none in that uninhabited island being to be had. It is no wonder that the sap of this tree should be so virulent, when rain or dew falling from its leaves on the naked body causes blisters on the skin, and even the effluvia of it are so noxious as to affect the senses of those which stand any time under its shade."

Oily substances are considered the best remedy for this poison. Some also recommend a large glass of sea-water to be drank instantly as a preventive.

The branches of the Manchineel are covered with a grayish, smooth bark. The leaves, which fall annually, are alternate, petiolate, numerous, oval, pointed, almost cordate at the base, slightly and distantly serrulate, dark green, rather thick, shining, veined, and transversely nerved, three to four inches long by about two inches wide. Stipules oval and caducous. The flowers are small and of a yellow color, monœcious, and grow upon straight, terminal spikes, like catkins. The male flowers are minute, collected together in clusters of about thirty together, each cluster subtended by a concave, caducous scale. The calycine scales are accompanied at their base by two large, lateral, orbicular, depressed glands. The fertile flowers are sessile and solitary. The drupe, in color and odor, is so like a small apple that it might easily be mistaken for it; it is shining, and of a yellowish-green color, with a white and milky pulp. It contains a thick, bony nut, full of angular crests which project almost through the skin; it has, ordinarily, six or seven, sometimes as many as fourteen? one-seeded cells, which have no spontaneous dehiscence or valves. The *male* flowers have a very small one-leaved, roundish, bifid calyx, with a straight, slender filament as long again as the calyx, bearing four roundish anthers. The *female* flower, like the preceding, has no corolla, and consists of a three-leaved calyx, with roundish, obtuse, connivent leaflets. The ovary is oval, superior, as long as the calyx, surmounted by a straight, short style, deeply divided into six or seven long, subulate, pointed, and reflected stigmas.

PLATE LX.

A branch of the natural size. a. The male flower. b. The apple-like drupe of the natural magnitude. c. A transverse section of the drupe having six cells and one abortive cell. d. The seed, of its natural magnitude. e. The kernel, with the inverted embryo of the natural size.

The poisonous Upas, (*Antiaris toxicaria,*) bearing solitary, female flowers with two styles and an unequal drupaceous fruit, though only of one cell, still approaches nearer to the anomalous Manchineel, in this family, than to any plant of the *Artocarpeæ*, with which it is so unnaturally associated.

Aleurites, by its fruit, a two-celled, two-seeded, indehiscent drupe, appears to be almost intermediate with Antiaris and Hippomane. We are unacquainted with the structure of the seed in Antiaris; but the obliquity of the fruit, and its swelling out more to one side, would seem to indicate the presence of two germs. These poisonous plants, as well as the Aleurites, seem to form a natural group, which further observation must decide; if so considered, they might bear the name of HIPPO-MANEÆ, from the well-known Manchineel, and will be distinguished chiefly from the EUPHORBIACEÆ by their indehiscent, drupaceous fruit of one or two to seven or more one-seeded cells, in place of three, the characteristic number in Euphorbiaceæ.

The large oily kernels of the *Aleurites triloba*, known in the Sandwich and Friendly Islands by the name of Too-tooe, are employed by the natives, generally, for lights: pierced with a skewer, they are lighted like a candle or a torch, and burn well and for a long time, giving out a bright flame and smoke. An excellent oil is obtained from these nuts by expression, which is used for a variety of purposes, and answers well for paint. It constitutes, likewise, one of the most ornamental and characteristic trees of the forest, visible at a great distance by the paleness and whiteness of its verdure, and hence the name of Aleurites given to it by Forster, from its mealy appearance. It grows rapidly and affords a fine shade, producing leaves which resemble those of the Plane Tree.

END OF VOL. I. OF NUTTALL.

STEREOTYPED BY L. JOHNSON & CO.
PHILADELPHIA.

www.ingramcontent.com/pod-product-compliance
Lightning Source LLC
Chambersburg PA
CBHW030738230426
43667CB00007B/762